THEMES OF THE

Times

ON

Astronomy

SECOND EDITION

A COLLECTION
OF ARTICLES FROM

The New York Times

PEARSON

Addison
Wesley

San Francisco   Boston   New York
Capetown   Hong Kong   London   Madrid   Mexico City
Montreal   Munich   Paris   Singapore   Sydney   Tokyo   Toronto

Editorial Director: Adam Black
Senior Acquisitions Editor: Lothlórien Homet
Project Editors: Deb Greco, Ashley Taylor Anderson
Editorial Assistant: Grace Joo
Managing Editor: Corinne Benson
Production Supervisor: Shannon Tozier
Manufacturing Manager: Pam Augspurger
Production Service and Composition: Carlisle Publishing Services
Text and Cover Printer: Courier, Stoughton

All pedagogical content, including article summaries, and "In Review" article-related questions test, was created by:

Doug Lombardi, University of Ariszona
Jeffrey Bennett, University of Colorado
Megan Donahue, Michigan State University
Nicholas Schneider, University of Colorado
Mark Voit, Michigan State University

ISBN 0-321-48756-7

All pedagogical content for this supplement has been created by Pearson Addison-Wesley, and includes: article summaries; cross-references to *The Essential Cosmic Perspective*, Fourth Edition, *The Cosmic Perspective*, Fourth Edition, and in the Table of Contents; and *Life in the Universe*, Second Edition, "In Review" questions that follow each article.

PEARSON

Addison
Wesley

1 2 3 4 5 6 7 8 9 10 -CRS- 09 08 07 06
www.aw-bc.com/physics

THEMES OF THE Times ON Astronomy

SECOND EDITION

Contents
to accompany chapters of *The Cosmic Perspective*, 4e

CHAPTER 1

9 Planets? 12? What's a Planet, Anyway?   1
October 4, 2005
By Dennis Overbye

CHAPTER 2

Venus Returns for Its Shining Hour   4
May 18, 2004
By Warren E. Leary

CHAPTER 3

Does Science Matter?   8
November 11, 2003
By William J. Broad and James Glanz

How Islam Won, and Lost, the Lead in Science   12
October 30, 2001
By Dennis Overbye

CHAPTER 4

Crunch! Oof! Well, That's Physics   17
November 16, 2004
By Henry Fountain

The Man Who Grasped The Heavens' Gravitas   21
October 8, 2004
By John Noble Wilford

Not Science Fiction: An Elevator to Space   24
September 23, 2003
By Kenneth Chang

CHAPTER 5

Explaining Ice: The Answers Are Slippery   27
February 21, 2006
By Kenneth Chang

CHAPTER 6

Mirror, Mirror   30
August 30, 2005
By Dennis Overbye

Telescopes of the World, Unite! A Cosmic Database Emerges   34
May 20, 2003
By Bruce Schechter

CHAPTER 7

Venus Returns for Its Shining Hour   4
May 18, 2004
By Warren E. Leary

NASA Planning Return to Moon Within 13 Years   37
September 20, 2005
By Warren E. Leary

The Allure of an Outpost on the Moon   40
January 13, 2004
By Kenneth Chang

CHAPTER 8

Second Disk Circling a Star May Provide Evidence of a Huge Hidden Planet   45
July 4, 2006
By John Noble Wilford

CHAPTER 9

Beyond Their Martian Dreams: Two Rovers Are Still Informing Experts Two Years Later   46
January 3, 2006
By Kenneth Chang

Mars' Round, Smooth Stones Have a Counterpart in
Utah   49
June 22, 2004
By Kenneth Chang

Scientists Report Evidence of Saltwater
Pools on Mars   51
March 24, 2004
By Warren E. Leary

Deadly and Yet Necessary, Quakes Renew
the Planet   74
January 11, 2005
By William J. Broad

CHAPTER 10

Methane in Martian Air Suggests Life Beneath the
Surface   53
November 23, 2004
By Kenneth Chang

Probes Reveal Methane Haze On a Dynamic Saturn
Moon   56
December 1, 2005
By Warren E. Leary

Studies Portray Tropical Arctic as Sultry
in Distant Past   77
June 1, 2006
By Andrew C. Revkin

Antarctica, Warming, Looks Ever More
Vulnerable   80
January 25, 2005
By Larry Rohter

CHAPTER 11

NASA Images Give New View of a Saturn Moon: No
Oceans, but a Sea of Sand   57
May 9, 2006
By Kenneth Chang

Space Probe Makes Science Fiction Wonders of
Childhood Real   59
January 25, 2005
By Lawrence M. Krauss

Saturn Moon Has Geysers, Hinting Life Is a
Possibility   61
March 10, 2006
By Kenneth Chang

Decoding the Dance of Saturn's Rings   63
July 6, 2004
By Kenneth Chang

CHAPTER 12

9 Planets? 12? What's a Planet, Anyway?   1
October 4, 2005
By Dennis Overbye

Vote Makes It Official: Pluto Isn't What
It Used to Be   64
August 25, 2006
By Dennis Overbye

Pluto's Exotic Playmates   67
September 12, 2006
By Kenneth Chang

NASA Launches Spacecraft On the First Mission to
Pluto   72
January 20, 2006
By Warren E. Leary

CHAPTER 13

Planet Group Similar to Solar System Is Found   42
May 18, 2006
By Dennis Overbye

Second Disk Circling a Star May Provide Evidence of a
Huge Hidden Planet   45
July 4, 2006
By John Noble Wilford

Search Finds Far-Off Planet Akin to Earth   70
January 26, 2006
By Dennis Overbye

CHAPTER 14

Someday the Sun Will Go Out and the World
Will End (but Don't Tell Anyone)   84
February 14, 2006
By Dennis Overbye

Tiny, Plentiful and Really Hard to Catch   86
April 26, 2005
By Kenneth Chang

CHAPTER 15

Stars on Diet: Weight Is Limited To 150 Suns,
Researchers Find   89
March 10, 2005
By Warren E. Leary

CHAPTER 16

Stars on Diet: Weight Is Limited To 150 Suns,
Researchers Find   89
March 10, 2005
By Warren E. Leary

Stardust Memories   91
May 5, 2006
By Frank Winkler

Seeing Mountains in Starry Clouds
of Creation   93
November 15, 2005
By Dennis Overbye

CHAPTER 17

Stardust Memories   91
May 5, 2006
By Frank Winkler

Life-or-Death Question: How
Supernovas Happen   95
November 9, 2004
By Dennis Overbye

CHAPTER 18

Dying Star Flares Up, Briefly Outshining Rest
of Galaxy   98
February 20, 2005
By Kenneth Chang

Astronomers Edging Closer To Gaining Black Hole Image   100
November 3, 2005
By Dennis Overbye

CHAPTER 19

In Galaxies Near and Far, New Views of Universe Emerge   102
January 14, 2003
By John Noble Wilford

CHAPTER 20

Three Dozen New Galaxies Are Found in Nearby Space   105
December 22, 2004
By Dennis Overbye

From Distant Galaxies, News of a "Stop-and-Go Universe"   107
June 3, 2003
By John Noble Wilford

CHAPTER 21

Milky Way And Neighbor Seen to Merge   109
January 10, 2006
By Warren E. Leary

Black Holes' Vast Power Is Documented   111
February 19, 2004
By John Noble Wilford

Music of the Heavens Turns Out to Sound a Lot Like a B Flat   113
September 16, 2003
By Dennis Overbye

CHAPTER 22

By X-Raying Galaxies, Researchers Offer New Evidence of Rapidly Expanding Universe   115
May 19, 2004
By Dennis Overbye

From Space, a New View of Doomsday   117
February 17, 2004
By Dennis Overbye

New Data on 2 Doomsday Ideas, Big Rip vs. Big Crunch   122
February 21, 2004
By James Glanz

CHAPTER 23

Astronomers Find the Earliest Signs Yet of a Violent Baby Universe   125
March 17, 2006
By Dennis Overbye

String Theory, at 20, Explains It All (or Not)   128
December 7, 2004
By Dennis Overbye

CHAPTER 24

Search for Life Out There Gains Respect, Bit by Bit   133
July 8, 2003
By Dennis Overbye

CHAPTER S1

Venus Returns for Its Shining Hour   4
May 18, 2004
By Warren E. Leary

THEMES OF THE

Times

ON
Astronomy

SECOND EDITION

Contents

to accompany chapters of *The Essential Cosmic Perspective*, 4e

CHAPTER 1

9 Planets? 12? What's a Planet, Anyway? 1
October 4, 2005
By Dennis Overbye

CHAPTER 2

Venus Returns for Its Shining Hour 4
May 18, 2004
By Warren E. Leary

CHAPTER 3

Does Science Matter? 8
November 11, 2003
By William J. Broad and James Glanz

How Islam Won, and Lost, the Lead in Science 12
October 30, 2001
By Dennis Overbye

CHAPTER 4

Crunch! Oof! Well, That's Physics 17
November 16, 2004
By Henry Fountain

The Man Who Grasped the Heavens' Gravitas 21
October 8, 2004
By John Noble Wilford

Not Science Fiction: An Elevator to Space 24
September 23, 2003
By Kenneth Chang

CHAPTER 5

Explaining Ice: The Answers Are Slippery 27
February 21, 2006
By Kenneth Chang

Mirror, Mirror 30
August 30, 2005
By Dennis Overbye

Telescopes of the World, Unite! A Cosmic Database Emerges 34
May 20, 2003
By Bruce Schechter

CHAPTER 6

NASA Planning Return to Moon Within 13 Years 37
September 20, 2005
By Warren E. Leary

The Allure of an Outpost on the Moon 40
January 13, 2004
By Kenneth Chang

Planet Group Similar to Solar System Is Found 42
May 18, 2006
By Dennis Overbye

Second Disk Circling a Star May Provide Evidence of a Huge Hidden Planet 45
July 4, 2006
By John Noble Wilford

Search Finds Far-Off Planet Akin to Earth 70
January 26, 2006
By Dennis Overbye

CHAPTER 7

Beyond Their Martian Dreams: Two Rovers Are Still Informing Experts Two Years Later 46
January 3, 2006
By Kenneth Chang

Mars' Round, Smooth Stones Have a Counterpart
in Utah   49
June 22, 2004
By Kenneth Chang

Scientists Report Evidence of Saltwater Pools
on Mars   51
March 24, 2004
By Warren E. Leary

Methane in Martian Air Suggests Life Beneath
the Surface   53
November 23, 2004
By Kenneth Chang

Deadly and Yet Necessary, Quakes Renew the
Planet   74
January 11, 2005
By William J. Broad

Studies Portray Tropical Arctic as Sultry in Distant
Past   77
June 1, 2006
By Andrew C. Revkin

Antarctica, Warming, Looks Ever More Vulnerable   80
January 25, 2005
By Larry Rohter

CHAPTER 8

Probes Reveal Methane Haze On a Dynamic Saturn
Moon   56
December 1, 2005
By Warren E. Leary

NASA Images Give New View of a Saturn Moon:
No Oceans, but a Sea of Sand   57
May 9, 2006
By Kenneth Chang

Space Probe Makes Science Fiction Wonders
of Childhood Real   59
January 25, 2005
By Lawrence M. Krauss

Saturn Moon Has Geysers, Hinting Life
Is a Possibility   61
March 10, 2006
By Kenneth Chang

Decoding the Dance of Saturn's Rings   63
July 6, 2004
By Kenneth Chang

CHAPTER 9

9 Planets? 12? What's a Planet, Anyway ?   1
October 4, 2005
By Dennis Overbye

Vote Makes It Official: Pluto Isn't What It Used
to Be   64
August 25, 2006
By Dennis Overbye

Pluto's Exotic Playmates   67
September 12, 2006
By Kenneth Chang

NASA Launches Spacecraft On the First Mission to
Pluto   72
January 20, 2006
By Warren E. Leary

CHAPTER 10

Someday the Sun Will Go Out and the World Will End
(but Don't Tell Anyone)   84
February 14, 2006
By Dennis Overbye

Tiny, Plentiful, and Really Hard to Catch   86
April 26, 2005
By Kenneth Chang

CHAPTER 11

Stars on Diet: Weight Is Limited to 150 Suns,
Researchers Find   89
March 10, 2005
By Warren E. Leary

Stardust Memories   91
May 5, 2006
By Frank Winkler

CHAPTER 12

Seeing Mountains in Starry Clouds of Creation   93
November 15, 2005
By Dennis Overbye

Life-or-Death Question: How Supernovas Happen   95
November 9, 2004
By Dennis Overbye

CHAPTER 13

Dying Star Flares Up, Briefly Outshining Rest
of Galaxy   98
February 20, 2005
By Kenneth Chang

Astronomers Edging Closer To Gaining Black Hole
Image   100
November 3, 2005
By Dennis Overbye

CHAPTER 14

In Galaxies Near and Far, New Views of Universe
Emerge   102
January 14, 2003
By John Noble Wilford

CHAPTER 15

Three Dozen New Galaxies Are Found in Nearby
Space   105
December 22, 2004
By Dennis Overbye

From Distant Galaxies, News of a "Stop-and-Go
Universe"   107
June 3, 2003
By John Noble Wilford

Milky Way And Neighbors Seen to Merge   109
January 10, 2006
By Warren E. Leary

Black Holes' Vast Power is Documented   111
February 19, 2004
By John Noble Wilford

Music of the Heavens Turns Out to Sound a Lot Like a B Flat   113
September 16, 2003
By Dennis Overbye

CHAPTER 16

By X-Raying Galaxies, Researchers Offer New Evidence of Rapidly Expanding Universe   115
May 19, 2004
By Dennis Overbye

From Space, a New View of Doomsday   117
February 17, 2004
By Dennis Overbye

New Data on 2 Doomsday Ideas, Big Rip vs. Big Crunch   122
February 21, 2004
By James Glanz

CHAPTER 17

Astronomers Find the Earliest Signs Yet of a Violent Baby Universe   125
March 17, 2006
By Dennis Overbye

String Theory, at 20, Explains It All (or Not)   128
December 7, 2004
By Dennis Overbye

CHAPTER 18

Search for Life Out There Gains Respect, Bit by Bit   133
July 8, 2003
By Dennis Overbye

THEMES OF THE
Times
ON
Astronomy

SECOND EDITION

Contents
to accompany chapters of *Life in the Universe*, 2e

CHAPTER 1

9 Planets? 12? What's a Planet, Anyway?　1
October 4, 2005
By Dennis Overbye

Does Science Matter ?　8
November 11, 2003
By William J. Broad and James Glanz

CHAPTER 2

How Islam Won, and Lost, the Lead in Science　12
Ocotber 30, 2001
By Dennis Overbye

Vote Makes It Official: Pluto Isn't What It Used to Be　64
August 25, 2006
By Dennis Overbye

CHAPTER 3

In Galaxies Near and Far, New Views of Universe Emerge　102
January 14, 2003
By John Noble Wilford

Three Dozen New Galaxies Are Found in Nearby Space　105
December 22, 2004
By Dennis Overbye

From Distant Galaxies, News of a "Stop-and-Go Universe"　107
June 3, 2003
By John Noble Wilford

Milky Way and Neighbors Seen to Merge　109
January 10, 2006
By Warren E. Leary

By X-Raying Galaxies, Researchers Offer New Evidence of Rapidly Expanding Universe　115
May 19, 2004
By Dennis Overbye

Astronomers Find the Earliest Signs Yet of Violent Baby Universe　125
March 17, 2006
By Dennis Overbye

CHAPTER 4

Deadly and Yet Necessary, Quakes Renew the Planet　74
January 11, 2005
By William J. Broad

Studies Portray Tropical Arctic as Sultry in Distant Past　77
June 1, 2006
By Andrew C. Revkin

Antarctica, Warming, Looks Ever More Vulnerable　80
January 25, 2005
By Larry Rohter

CHAPTER 6

Search for Life Out There Gains Respect, Bit by Bit　133
July 8, 2003
By Dennis Overbye

CHAPTER 8

Beyond Their Martian Dreams: Two Rovers Are Still Informing Experts Two Years Later　46
January 3, 2006
By Kenneth Chang

Mars' Round, Smooth Stones Have a Counterpart in Utah　49
June 22, 2004
By Kenneth Chang

Scientists Report Evidence of Saltwater Pools
on Mars   51
March 24, 2004
By Warren E. Leary

Methane in Martian Air Suggests Life Beneath
the Surface   53
November 23, 2004
By Kenneth Chang

CHAPTER 9

Probes Reveal Methane Haze On a Dynamic
Saturn Moon   56
December 1, 2005
By Warren E. Leary

NASA Images Give New View of a Saturn Moon:
No Oceans, but a Sea of Sand   57
May 9, 2006
By Kenneth Chang

Space Probe Makes Science Fiction Wonders
of Childhood Real   59
January 25, 2005
By Lawrence M. Krauss

Saturn Moon Has Geysers, Hinting Life
Is a Possibility   61
March 10, 2006
By Kenneth Chang

CHAPTER 11

Planet Group Similar to Solar System Is Found   42
May 18, 2006
By Dennis Overbye

Second Disk Circling a Star May Provide Evidence
of a Huge Hidden Plant   45
July 4, 2006
By John Noble Wilford

Search Finds Far-Off Planet Akin to Earth   70
January 26, 2006
By Dennis Overbye

CHAPTER 12

Search for Life Out There Gains Respect, Bit by
Bit   133
July 8, 2003
By Dennis Overbye

As discussed in the Special Topic box in Chapter 1 of your text — and in later articles in this *Themes of the Times* supplement — Pluto was recently demoted from being one of the official planets to being a "dwarf planet." This article, written less than a year before the official decision was made, summarizes some of the key factors that played into the debate.

Note: The object referred to in this article as "Xena" of "103 UB313" is now officialy named Eris; its moon is named Dysnomia.

9 Planets? 12? What's a Planet, Anyway?

By Dennis Overbye
The New York Times, October 4, 2005

In my daughter's circle of friends, one 3-year-old named Jared can reel off the names of all the planets. He and his parents are pretty proud, justly in my estimation, of this achievement. Little does he know, however, that the lords of astronomy are working against him.

For the last 18 months, a committee appointed by the International Astronomical Union has been pondering in frustrating exactitude whether the word "planet" means anything anymore.

Last month Nature reported that the committee was ready to propose dumping the bare term "planet" in favor of an expanded, more embellished set of terms like "terrestrial planets," "transNeptunian planets" and so forth.

But that turned out to be a false alarm, according to the committee's chairman, Iwan Williams of Queen Mary College in London. He said in an e-mail message that although a majority favored the redefinition many other ideas were blooming and contending so fractiously that he despaired of ever reaching general agreement.

Dr. Williams said, "Up to this point I have been hoping for a consensus, but I guess we might need to go for a majority vote."

The solar system is much more complicated now, astronomers say, than in 1930 when Clyde Tombaugh added Pluto to the inventory of wandering lights circling the Sun. In addition to Earth, Mars, Venus, Jupiter, Saturn, Mercury, Neptune, Uranus and Pluto, schoolchildren now learn that there are also comets and asteroids bumping about in the night.

But there are also the Oort cloud, a hypothesized halo of cometary bits hibernating in deep, deep space, and the Kuiper Belt, a ring of icy bodies beyond Neptune's orbit. Not to mention the dozens of moons circling the planets.

Pluto is the big problem. Is it a planet or not? Some astronomers have long argued that its small size, less than one-fifth the diameter of Earth, and a weird tilted orbit that takes it inside Neptune every couple hundred years make Pluto more like a Kuiper Belt body than a full-fledged planet.

A furor arose five years ago when my colleague Kenneth Chang reported that the new Rose Center for Earth and Space at the American Museum of Natural History in New York had demoted Pluto, calling it a "Kuiper Belt object" rather than a planet.

The controversy became more desperate this summer when astronomers discovered a new object larger than Pluto orbiting in the Kuiper Belt at a distance of nine billion miles from the Sun. Michael E. Brown of the California Institute of Technology, its discoverer, has said it will be fine with him if Pluto is demoted to a minor planet, but, he argues, if Pluto is a planet, so is the new object, which he nicknamed Xena, making it the 10th planet. Last Friday Dr. Brown announced that Xena has a tiny moon, making it seem even more planetlike.

Brian Marsden, an astronomer at the Harvard-Smithsonian Center for Astrophysics, directs the astronomical union's Minor Planet Center, a clearinghouse for solar system discoveries, thinks that both Pluto and Dr. Brown's Xena should be called minor planets. He is one of those who support the idea of categorizing planets into groups. But according to Dr. Williams, other members of the panel have championed other ideas, for example, that planets should be larger than 2,000 kilometers (or about 1,250 miles) in diameter (Pluto is about 1,500 miles).

Dr. Neil DeGrasse Tyson, of the Rose Center, who is not a member of the astronomical union committee, said the proposed naming scheme sounded a lot like the museum's system. He said, however, that the need to assign adjectives to the word "planet" might mean it was time to retire the term altogether.

Asked what he would replace it with, Dr. Tyson said he hoped the geologists could come up with something and offered up words like "terrestrials," for balls of dirt and rock like Earth; "Jovians," for giant gaseous planets like Jupiter and Saturn; comets; and so forth.

Not only did the panel members disagree on the definition of a planet, at last report they could not even agree, it seemed, on whether they were making progress. Within the space of a few minutes the other week, I received one e-mail message from Dr. Marsden saying he was optimistic and another from Alan Boss of the Carnegie Institution of Washington complaining that his morning e-mail gave him no sense that they were close to bringing the issue to a close.

In another e-mail message, Dr. Boss described the process as "like trying to shovel frogs into a wheelbarrow — they keep jumping out again."

The new object, now known poetically as 2003 UB313, is destined to languish nameless until the astronomical union panel comes to a conclusion and thus decides which part of the astronomical bureaucracy is responsible for major planets or minor planets, will have to approve a new name.

"Every time it looks as though we might be approaching a consensus, rather severe disagreement has a way of breaking out again," Dr. Marsden said. "It is all very unfortunate."

Even the idea that a majority of the committee was leaning anywhere was hotly disputed by S. Alan Stern of the Southwest Research Institute in Boulder, Colo., who is the principal investigator of a coming National Aeronautics and Space Administration mission to Pluto and the Kuiper Belt.

Dr. Stern has suggested that the criterion of planethood be roundness — a body big enough for gravity to have conquered geological and mechanical forces. That would include in the roll call of planets not only Pluto, but dozens of objects he thinks are yet to be discovered out in the Kuiper Belt.

"I think it'll be obvious very shortly that dwarf planets surely outnumber the planets our fathers' generation thought of as the mainstream," Dr. Stern, 47, said. If it should turn out in the years to come that Earth is not actually a very good example of a planet, that would not be the first instance in which astronomers have painted themselves into a corner, linguistically.

Being limited to looking in the practice of their science, astronomers tend to characterize and classify new phenomena by their appearances, their colors, for example. Objects that resemble one another then get lumped together under the name of their progenitor until enough differences accumulate to start a new category.

So it was that a starlike object known as BL Lacertae, the progenitor of a kind of exploding galaxy, was eventually said to not be a good example of a BL Lacertae object.

Astronomers are not alone in having a silly name problem, but they may be alone in agonizing so much about it. Physicists have shown little restraint in creating names like neutrinos, quarks, squarks, gluons, photinos, selectrons, strangeness and charm, not to mention strings and branes.

No commission regulates what physicists call things. "Basically physicists are too undisciplined to let anyone else tell us what to name something," said Gordon Kane of the University of Michigan. "It's mainly whatever name catches on."

Dr. Williams of the astronomical union said there was no deadline for his committee's decision on planetary nomenclature. The urgency, he said, comes from the need to find a way of giving Dr. Brown's Xena an official name. Early this week Dr. Williams circulated a new proposal, defining anything with a radius of more than 1,000 kilometers (roughly Pluto) as a planet, asking for a vote within two weeks. So far there are no indications of a groundswell.

Dr. Boss of the Carnegie Institution, who favors and hopes that the adjective-planet compromise will catch on in astronomy, said he thought the public would be "thrilled" to realize that astronomy had progressed so far as to require a re-sorting of the primary components of the solar system. "Science marches on," he said, "and this brouhaha is a sterling reminder of both the joy and the pain of this process."

I cannot argue with his logic and his desire for clarity. But at the risk of being a curmudgeon, and in the interest of 3-year-olds everywhere who are reaching for a basic comprehension of how the universe is put together and where they stand, I think the astronomers should take a page from the lawyers and jurists we're hearing so much from these days.

I think they should adopt "stare decisis," Latin for "stand by that decided."

By precedent Pluto is a planet. If we agree on that, then we can look forward to many more planets.

There were smiles all around the office here when the putative 10th planet was announced last July. There is something ennobling and hopeful about

living at a time when a 10th planet is added to the solar system, and maybe an 11th and 12th as planned surveys of the trans-Neptunian void are carried out. It is like being young again and finding out your family is larger than you thought.

It is like living in an expanding universe.

Like a Smaller Earth-Moon System, 10th Planet Has Its Own Orbiter

The 10th planet has a moon.

Michael E. Brown, a professor of planetary astronomy at the California Institute of Technology, who discovered the planet, 2003 UB313, an object in the outer solar system bigger than Pluto, announced on Friday that it has a small moon circling it.

Since Dr. Brown gave the planet the unofficial nickname Xena, after the main character of a television series about a Greek warrior princess, he called the moon Gabrielle, after Xena's sidekick on the show. Dr. Brown has not yet divulged his proposal for the official name of 2003 UB313.

Gabrielle, one one-hundredth as bright as the planet it circles, was spotted in a photograph taken Sept. 10 at the Keck Observatory in Hawaii. More observations are needed to pin down its orbit. Dr. Brown says that it is about 25,000 miles from Xena and takes one to three weeks to complete an orbit.

"The whole thing is like the Earth-Moon system scaled down by a factor of 5 or 10," Dr. Brown said. "It's kind of a cool story."

The findings have been submitted to Astrophysical Journal Letters.

A precise determination of Gabrielle's orbit will allow astronomers to calculate the mass of the 10th planet. Moons apparently are not rare in the Kuiper Belt, a ring of debris beyond Neptune. Astronomers have now found moons around three of the largest Kuiper Belt objects: Pluto, the 10th planet, and 2003 EL61, another large body spotted by Dr. Brown.

That suggests Pluto is not as much an oddball as once thought. "You are finally learning a little more what's going on out there," Dr. Brown said.

IN REVIEW

1. Why was Pluto's planetary status controversial even before the 2006 decision to demote it from the list of planets?

2. Where are Pluto and "Xena" (Eris) located in our solar system?

3. Based on the discussions in this article, summarize some of the possible options that astronomers considered for resolving the debate about the number of planets in our solar system.

4. The article's author argued for keeping Pluto a planet by the doctrine of "stare decisis." What did he mean? Do you agree with his rationale? Why or why not?

5. If it had been up to you, which option would you have chosen for defining "planet"? With your definition, how many planets would there be in our solar system? Explain.

With the exception of the Moon, the planet Venus is the brightest object in the night sky. Shining brightly just after sunset or before sunrise, Venus has captured the imagination of sky watchers since ancient times. As a planet, Venus wanders across the sky in a manner much different than the stars, and occasionally Venus appears to move across the Sun. Because Venus is much more distant from the Earth than the Moon is, when its path transverses the solar disk it does not fully eclipse the light, but rather is seen (through special filtered safety glass) as a small blemish. This movement, called a solar transit, is more than a fanciful curiosity. Such transits of Venus have helped scientists accurately determine planetary distances.

Venus Returns for Its Shining Hour

By Warren E. Leary
The New York Times, **May 18, 2004**

The world is about to witness a rare spectacle that once launched expeditions to ideal vantage points around the globe and inspired millions of people to venture outside and stare at the heavens.

On June 8, people in the right places on Earth will be able to see Venus move across the face of the Sun in a kind of minieclipse that is visible twice every century or so. The last such occurrence, called a transit of Venus, was in 1882. It inspired an international effort to use the event to answer one of the most pressing scientific questions of the day: What is the exact distance between the Sun and Earth?

Although studies of the event failed to provide an exact answer, they did narrow the range of estimates and measurements, and ushered in an era of investing in science as a symbol of national prestige. For the last event, the United States government mustered eight expeditions to make observations around the world, partly because Britain, France, Russia and other rivals did the same.

By bouncing radar signals off the Sun and Venus and using spacecraft measurements, scientists in the 1960's calculated that the average Sun-to-Earth distance is 92,955,859 miles, a measure called the astronomical unit.

Scientists realized for centuries that if they could find out that number, they could use the formulas of the 17th-century astronomer Johannes Kepler to calculate the size of the solar system and the exact distances between the planets.

"This was the most important question of its day in astronomy," said Dr. Jay M. Pasachoff, a professor of astronomy at Williams College. "And using the transits of Venus to calculate the astronomical unit was the best way to do it."

Although transits of Venus have occurred for thousands of years, the first report of its subtle crossing of the Sun was in 1639. The transits occur when the orbits of Venus, Earth and the Sun put them into alignment along the same plane.

Since 1639, transits have occurred in 1761, 1769, 1874 and 1882. If someone misses the one next month, there will be another opportunity on June 6, 2012. After that, more than a century will pass before the next transits, in 2117 and 2125. Because of its rarity, the transit next month, best viewed from Europe and the Mideast, is generating great scientific and public interest, said Dr. Steven J. Dick, chief historian for the National Aeronautics and Space Administration. Dr. Dick has written extensively on the 18th- and 19th-century transits.

No one alive today saw the last transit, he said, and seeing the next two will be the only chance most people have.

"These are truly once-in-a-lifetime events," Dr. Dick said. "Although the scientific importance has diminished, I think there will be a lot of interest this time among the public, based on e-mail I've seen from around the world."

Dr. David DeVorkin, curator of the history of astronomy at the National Air and Space Museum, said the 1874 and 1882 transits were prominently featured in newspapers and magazines. A carnival atmosphere pervaded Wall Street for the transit on Dec. 6, 1882, with people crowding the area and staring up through smoked glass.

"It was a popular diversion," Dr. DeVorkin said. "Something maybe everybody didn't try to see, but everybody talked about it."

Scientific interest persists. Instruments aboard at least three Sun-watching satellites, as well as ground telescopes, will follow the event. Researchers will use Venus' trek to test techniques and instruments that can be used to detect planets in other solar systems.

More than 120 extrasolar planets have been discovered orbiting other stars, most of them huge bodies found because their gravity affected the motion of their stars.

Astronomers have recently detected a small number of far planets by measuring the fluctuations that they cause in light from the stars they circle. In 2007, NASA plans to launch the Kepler spacecraft to monitor Sun-like stars in hope of detecting Earth-size planets through small decreases in star brightness.

Although denied a direct view of the transit because it occurs at night in the American West, astronomers with the University of Arizona hope to get an indirect view. Dr. Glenn H. Schneider

A Rare Trip Past the Sun

On June 8, Venus will track across the disk of the Sun for the first time since 1882. Whether you can view it will depend on where you are.

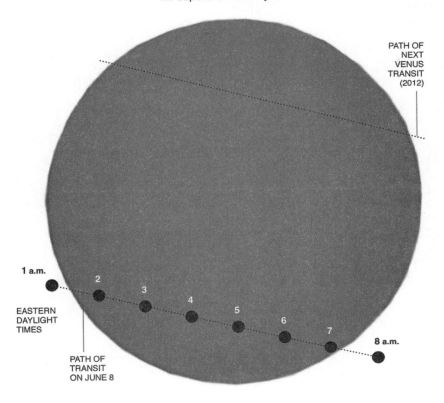

PATH OF NEXT VENUS TRANSIT (2012)

1 a.m.

EASTERN DAYLIGHT TIMES

PATH OF TRANSIT ON JUNE 8

8 a.m.

A TOOL FOR EARLY ASTRONOMERS Scientists estimated the distance between the Earth and Sun during a transit of Venus this way:

1 Two observers with a great distance between them — one in Alaska, for example, and one in Argentina — observe the transit.

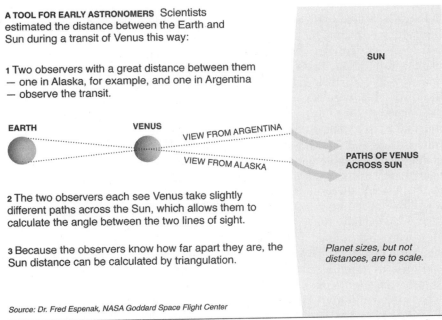

EARTH

VENUS

VIEW FROM ARGENTINA

VIEW FROM ALASKA

SUN

PATHS OF VENUS ACROSS SUN

Planet sizes, but not distances, are to scale.

2 The two observers each see Venus take slightly different paths across the Sun, which allows them to calculate the angle between the two lines of sight.

3 Because the observers know how far apart they are, the Sun distance can be calculated by triangulation.

Source: Dr. Fred Espenak, NASA Goddard Space Flight Center

The New York Times; Sun H-alpha image by University Corporation for Atmospheric Research

said he and a colleague, Paul S. Smith, would try to use the Steward Observatory in Tucson to measure about a half-hour of sunlight from the end of the transit as it reflects off the Moon.

"We want to see if we can detect the signature of Venus' atmosphere spectroscopically from sunlight reflecting off the moon, as if it was a reading coming from a faraway star," Dr. Schneider said.

The transits generally occur in a predictable pattern of two occurring in an eight-year period, followed by one 105 1/2 years later and another eight years after that. After an additional 121 1/2 years, the pattern repeats. The paired eight-year sightings occur because a Venusian year equals 224.7 Earth days, making 13 Venusian years equal to eight Earth years.

That allows the planets to return to about the same alignment with the Sun they had been in eight years earlier, after which they go out of sync for more than a century.

On Tuesday, June 8, observers lucky enough to view the entire transit will see Venus as a small black spot crossing the southern hemisphere of the Sun from left to right. The planet, entering the disc of the Sun at the 8 o'clock position, will take six hours to cross the bright face before exiting at the 5 o'clock position.

Venus, appearing as a round black dot with a diameter one thirty-second of the Sun's, is widely expected to cause a one-tenth of 1 percent drop in sunlight that reaches Earth during the event.

Location is everything, particularly when trying to witness celestial events. The entire transit will be visible in Europe, most of Africa, the Mideast and most of Asia. The unlucky regions of the globe where the event occurs at night, and is unviewable, include western North America, including most of the United States west of the Rockies; southern Chile and Argentina; Hawaii; and New Zealand.

Some regions will see just part of the transit, because the Sun sets while it is in progress. Those areas include Australia, Indonesia, Japan, the Philippines, Korea and Southeast Asia.

Likewise, the Sun rises with the transit in progress over eastern North America, the Caribbean, western Africa and most of South America, allowing observers a brief view before the event ends. How much early risers see will depend on the weather and how high the Sun rises above the horizons before Venus moves out of view.

In New York, sunrise will be at 5:25 a.m., and Venus is to begin exiting the solar disc at 7:06, when the Sun is 17 degrees above the horizon. The planet's final contact with the edge of the Sun should occur at 7:26 a.m., when the Sun is 20 degrees high. Times are similar for most cities in the Eastern time zone and one hour earlier in the Central time zone. But moving West means that the Sun is lower on the horizon.

Modern interest in planetary transits can be traced from Kepler. Based on his calculations of planetary motion, he wrote in 1627 that Mercury would cross the face of the Sun in November 1631 and that Venus would follow on Dec. 6 that year. Kepler suggested that observers placed at widely different points on Earth could indirectly calculate the distance to the Sun by using Venus.

Knowing the distance between observers and the different angles from which they viewed the transit, astronomers could calculate the distance to Venus and use that to compute the Earth-to-Sun measurement, he reasoned.

Kepler died the year before the 1631 Venus transit, but he would not have seen it had he lived, because it occurred at night in Europe. He would have also missed the next transit, in 1639, because he made a miscalculation that failed to predict it.

Fortunately, a young English astronomer, Jeremiah Horrocks, became interested in Kepler's work and, in recalculating some of the German's tables, discovered that a transit would occur on Nov. 24, 1639. Horrocks witnessed part of the transit from his home in Much Hoole, Lancashire, and a friend whom he notified by letter, William Crabtree, saw it from Manchester.

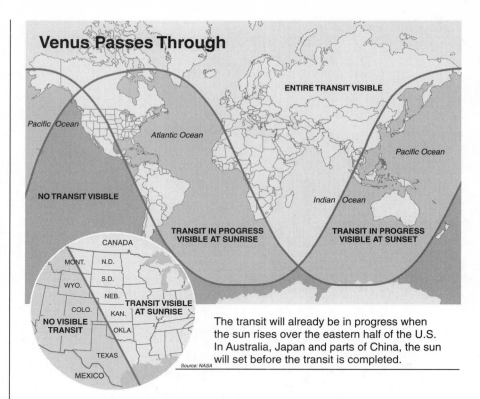

Venus Passes Through

ENTIRE TRANSIT VISIBLE

Pacific Ocean

Atlantic Ocean

Pacific Ocean

NO TRANSIT VISIBLE

Indian Ocean

TRANSIT IN PROGRESS VISIBLE AT SUNRISE

TRANSIT IN PROGRESS VISIBLE AT SUNSET

CANADA

MONT. N.D.

S.D.

WYO. NEB.

COLO. KAN. TRANSIT VISIBLE AT SUNRISE

NO VISIBLE TRANSIT OKLA.

TEXAS

MEXICO

Source: NASA

The transit will already be in progress when the sun rises over the eastern half of the U.S. In Australia, Japan and parts of China, the sun will set before the transit is completed.

The next transits, in the 18th century, drew much more attention, thanks to Edmond Halley, the British astronomer best known for the comet that bears his name. Halley suggested using the 1761 and 1769 transits to calculate the Sun-to-Earth distance by having observers time the events from widely spaced latitudes and trace the planet's path across the Sun's face as they saw it from their positions. By measuring the angular shifts of the paths based on the timings, Halley reasoned, the astronomical unit could be calculated.

Although Halley died in 1742, his plan guided many observations made of the two transits from around the world. But the results varied widely and were disappointing. Among those trying to work on the problem in 1769 was the British explorer Capt. James Cook, who took his ship, the Endeavour, on its first voyage to the South Pacific to observe the transit from Tahiti.

Cook and others were frustrated in their observations by the inability to time the exact moment when the edges of the planet and the Sun appeared to touch. When Venus nears the edge of the disc of the Sun, its black circle appears to ooze toward the edge of the sun without showing a clear point of contact. Although the precise second of contact was needed for calculations, this so-called "black drop" phenomenon caused observers watching the same event to disagree by several seconds up to a minute on when the outer edges touched.

Cook and other observers speculated that the problem was the distortion of light through the Venusian atmosphere.

Earlier this year, using spacecraft observations, Dr. Pasachoff and other scientists concluded that the black drop effect was caused by a combination of images' blurring in small-aperture telescopes and the natural dimming of sunlight near the Sun's visible edge.

In the 19th-century transits, scientists tried to overcome that effect and other imperfections with better telescopes and the introduction of photography. Still measuring and timing transits never led to finding the precise Sun-to-Earth distance.

William Harkness of the United States Naval Observatory refined results from the 1882 transits and in 1894 came up with an astronomical unit measure of 92,797,000 miles. But the work of another Naval Observatory scientist, Simon Newcomb, was adopted as the world standard at a 1896 meeting in Paris, Dr. Dick said. Newcomb, who gave little credence to transit data, combined values from several sources including speed-of-light star readings, to come up with a figure of about 92,872,000 miles. Both were close to the modern value of 92,955,859 miles, but precision is critical in astronomical terms.

Nevertheless, Dr. Dick said, the transits of Venus remain important because the desire to define the astronomical unit—and to maintain or gain scientific prestige—led many nations to mount competing expeditions. In 1874, Russian sent out 26 expeditions, Britain 12, the United States 8, Germany and

Photo by U.S. Naval Observatory

Testing the equipment at the 1874 American transit-of-Venus observation station on Kerguelen Island. On June 8, and for the first time since 1882, Earthlings will stare at the heavens with a chance to see Venus move across the face of the Sun.

France 6 each, Italy 3 and the Netherlands 1.

"You could compare it with the space race in the 20th century," he said.

How to Watch Without Harm

When viewing solar events like eclipses or the transit of Venus, precautions are needed. Never look directly at the Sun. The direct gaze can lead to severe eye damage or blindness, experts say.

Sunglasses and clouds do not protect the eyes, and viewing the Sun through unfiltered telescopes, binoculars or cameras can result in instant and permanent damage. Telescopes and binoculars should be equipped with special undamaged solar filters. Glasses with solar lenses are available commercially, but even then do not stare at the Sun for long periods.

The transit of Venus can safely be seen if viewed indirectly, using telescopes or pinhole boxes to focus the image on a screen or paper opposite the Sun.

Information on viewing solar events is at these Web sites: www.transitofvenus.org/safety./gov/eclipse/SEhelp/safety2.html

More Web resources can be found at: nytimes.com/science.

IN REVIEW

1. In what ways did the 1882 Venus transit usher in a new era of using science to bolster national prestige?

2. Why would two transits that are just seven years apart be advantageous for calibrating the distance between the Earth and the Sun compared to transits that are spaced more than one hundred years apart?

3. We can now accurately measure the distance from the Earth to the Sun using radar. But the Venus transit still provides some useful information to astronomers. List some different types of information that scientists get from the transit.

4. Before radar was invented, how did astronomers use transits to calculate the distance from the Earth to the Sun?

5. Why would Mercury and Venus be the only transits visible from Earth? If you were standing on the surface of Saturn's moon Titan, what planetary transits might you be able to see?

In recent years there has been an increasing divide between those who trust and those who are suspicious of science. A lack of understanding of the basic characteristics or "hallmarks" of science may be partially to blame. Science requires continual questioning and testing of ideas, and without a fundamental understanding of how observations and models are interweaved to form scientific theories, the process of science may seem uncertain and unreliable. Also, technology, acting as the interface between science and society, while providing many benefits, can have negative consequences. Some people blame science for causing these technological problems.

Does Science Matter?

By William J. Broad and James Glanz
The New York Times, **November 11, 2003**

Through its rituals of discovery, science has extended life, conquered disease and offered new sexual and commercial freedoms. It has pushed aside demigods and demons and revealed a cosmos more intricate and awesome than anything produced by pure imagination.

But there are new troubles in the peculiar form of paradise that science has created, as well as new questions about whether it has the popular support to meet the future challenges of disease, pollution, security, energy, education, food, water and urban sprawl.

The public seems increasingly intolerant of grand, technical fixes, even while it hungers for new gadgets and drugs. It has also come to fear the potential consequences of unfettered science and technology in areas like genetic engineering, germ warfare, global warming, nuclear power and the proliferation of nuclear arms.

Tension between science and the public has thrown up new barriers to research involving deadly pathogens, stem cells and human cloning. Some of the doubts about science began with the environmental movement of the 1960's.

"The bloom has been coming off the rose since 'Silent Spring,'" said Dr. John H. Gibbons, President Bill Clinton's science adviser, of Rachel Carson's 1962 book on the ravages of DDT. Until then, he said, "People thought of science as a cornucopia of goodies. Now they have to choose between good and bad."

"The urgency," he said, "is to reestablish the fundamental position that science plays in helping devise uses of knowledge to resolve social ills. I hope rationality will triumph. But you can't count on it. As President Chirac said, we've lost the primacy of reason."

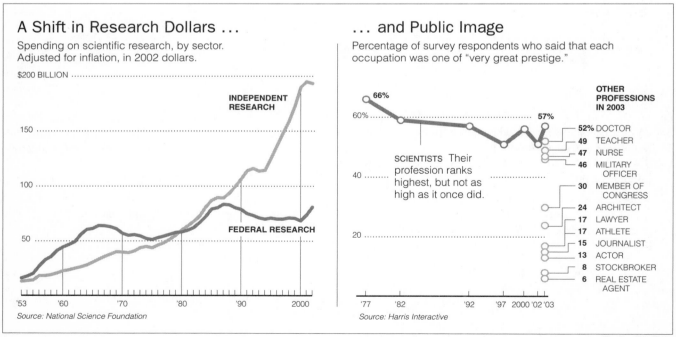

A Shift in Research Dollars ...
Spending on scientific research, by sector. Adjusted for inflation, in 2002 dollars.

Source: National Science Foundation

... and Public Image
Percentage of survey respondents who said that each occupation was one of "very great prestige."

Source: Harris Interactive

The New York Times

Science has also provoked a deeper unease by disturbing traditional beliefs. Some scientists, stunned by the increasing vigor of fundamentalist religion worldwide, wonder if old certainties have rushed into a sort of vacuum left by the inconclusiveness of science on the big issues of everyday life.

"Isn't it incredible that you have so much fundamentalism, retreating back to so much ignorance?" remarked Dr. George A. Keyworth II, President Ronald Reagan's science adviser.

The disaffection can be gauged in recent opinion surveys. Last month, a Harris poll found that the percentage of Americans who saw scientists as having "very great prestige" had declined nine percentage points in the last quarter-century, down to 57 from 66 percent. Another recent Harris poll found that most Americans believe in miracles, while half believe in ghosts and a third in astrology—hardly an endorsement of scientific rationality.

"There's obviously a kind of national split personality about these things," said Dr. Owen Gingerich, a historian of astronomy at the Harvard-Smithsonian Center for Astrophysics who speaks often of his Christian faith.

"Science gives you very cold comfort at times of death or sickness or so on," Dr. Gingerich said.

In this atmosphere of ambivalence, research priorities have become increasingly politicized, some scientists say.

"Right now it's about as bad as I've known it," said Dr. Sidney Drell, a Stanford University physicist who has advised the federal government on national security issues for more than 40 years.

As the world marches into a century born amid fundamentalist strife in oil-producing nations, a divisive political climate in the United States and abroad and ever more sophisticated challenges to scientific credos like Darwin's theory of evolution, it seems warranted to ask a question that runs counter to centuries of Western thought: Does science matter? Do people care about it anymore?

The Context

Breakthroughs And Disenchantment

Clearly, science has mattered a lot, for a long time. Advances in food, public health and medicine helped raise life expectancy in the United States in the past century from roughly 50 to 80 years. So too, world population between 1950 and 1990 more than doubled, now exceeding six billion. Biology discovered the structure of DNA, made test-tube babies and cured diseases. And the decoding of the human genome is leading scientists toward a detailed understanding of how the body works, offering the hope of new treatments for cancer and other diseases.

"For a lot of people, life has gotten better," said Dr. James D. Watson, co-discoverer of the double helix. "You don't know what it was like in 1950. It wasn't just the dreariness of Bing Crosby that made life tough."

In physics, breakthroughs produced digital electronics and subatomic discoveries. American rocket science won the space race, put men on the moon, probed distant planets and lofted hundreds of satellites, including the Hubble Space Telescope.

But major problems also arose: acid rain, environmental toxins, the Bhopal chemical disaster, nuclear waste, global warming, the ozone hole, fears over genetically modified food and the fiery destruction of two space shuttles, not to mention the curse of junk e-mail. Such troubles have helped feed social disenchantment with science.

When the cold war ended, the physical sciences began to lose luster and funding. After spending $2 billion, Congress killed physicists' preeminent endeavor, the Superconducting Super Collider, an enormous particle accelerator.

"Suddenly, Congress wasn't interested in science anymore," said Fred Jerome, a science policy analyst at the New School.

At the same time, industry spending on research soared to twice that of the federal government, about $180 billion last year, according to the National Science Foundation. One result is that Americans see more drugs, cellphones, advanced toys, innovative cars and engineered foods and less news about the fundamental building blocks and great shadowy vistas of the universe.

The main exceptions to the downward trend in the federal science budget are for health and weapons. This year, spending on military research hit $58 billion, higher in fixed dollars than during the cold war.

Meanwhile, other countries are spending more on research, taking some of the glory that America once monopolized. Japan, Taiwan and South Korea now account for more than a quarter of all American industrial patents, according to CHI Research. Europe is working on what will be the world's most powerful atom smasher. The British are now flying the first probe in a quarter century to look for evidence of life on Mars.

The Contradictions

New Challenges, But Also Threats

Despite the explosion in the life sciences, cancer still darkens many lives, and the flowering of biotechnology has fed worries about genetically modified foods and organisms as well as the pending reinvention of what it means to be human. Many people worry that the growing power of genetics will sully the sanctity of human life.

Last month, the President's Council on Bioethics issued a report warning that biotechnology in pursuit of human perfection could lead to unintended and destructive ends. Experts also worry about terrorists using advances in biology for intentional harm, perhaps on vast new scales.

"As this becomes ever easier and cheaper, it's only a matter of time before some misguided people decide to infect the world," said Dr. Philip Kitcher, a philosopher of science at Columbia University. Last month, a panel of the National Academy of Sciences recommended wide review of experiments that could lead to biological weapons.

The physical sciences seem to have lost what was once a good story line. Without the space race and the cold war, and perhaps facing intrinsic limits as well as declining budgets, they are slightly adrift. Some observers worry that physics has entered a phase of diminishing returns. That theme runs through "The End of Science," a 1997 book by John Horgan.

In an interview, Mr. Horgan noted that physicists no longer make nuclear arms and have lost momentum on taming fusion energy, which powers the sun, and on developing a theory of everything, a kind of mathematical glue that would unite the sciences. Abstract physics, he said, "has wandered off into the fantasy land of higher dimensions and superstring theory and has really lost touch with reality."

Other experts disagree, noting that scientific fields rise and fall in cycles and that physics may be poised for new strides. "You can smell discovery in the air," said Dr. Leon M. Lederman, a Nobel laureate in physics and an architect of the supercollider. "The sense of imminent revolution is very strong."

Despite the decline in prestige recorded in the recent Harris poll, scientists still top the list of 22 professions in terms of high status, ahead of doctors, teachers, lawyers and athletes.

"Science is one of the charismatic activities," said Dr. Gerald Holton, a professor of physics and the history of science at Harvard. "This keeps our interest in science at some level even if we are deeply troubled by some aspects of its technical misuse."

Polls by the National Science Foundation perennially identify contradictions. Its latest numbers show that 90 percent of adult Americans say they are very or moderately interested in science discoveries. Even so, only half the survey respondents knew that the Earth takes a year to go around the Sun.

"The easy answer is, 'Oh, I'm interested,'" said Melissa Pollak, a senior analyst at the National Science Foundation. "I'm not quite sure I believe those responses."

The Competition

The Battles Increase Over Darwin's Theory

A simple number jars many scientists: about two-thirds of the public believe that alternatives to Darwin's theory of evolution should be taught in public schools alongside this bedrock concept of biology itself.

The organized opposition to the mainstream theory of evolution has become vastly more sophisticated and influential than it was, say, 25 years ago. The leading foes of Darwin espouse a theory called "intelligent design," which holds that purely random natural processes could never have produced humans. These foes are led by a relatively small group of people with various academic and professional credentials, including some with advanced degrees in science and even university professorships.

Backers of intelligent design say they are simply pointing up shortcomings in Darwin's theory. Scientists have publicly rallied in response, last week staving off an effort at the Texas State Board of Education to have intelligent design taught alongside evolution.

"It just absolutely boggles the mind," said Dr. James Langer, a physicist at the University of California at Santa Barbara who is vice president of the National Academy of Sciences. "I wouldn't want my doctor thinking that intelligent design was an equally plausible hypothesis to evolution any more than I would want my airplane pilot believing in the flat Earth."

Science has, in fact, sold itself from the start as something more than a utilitarian exercise in developing technologies and medicines. Einstein—who often used religious and philosophical language to explain his discoveries—seemed to tell humanity something fundamental about the fabric of existence. More recently, the cosmologist Stephen Hawking said that discovering a better theory of gravitation would be like seeing into "the mind of God."

Such rhetorical flourishes are as much derided as admired by the bulk of working scientists, who as a culture have drifted closer to the thinking of Steven Weinberg, another Nobel Prize winner in particle physics, who famously wrote that "the more the universe seems comprehensible, the more it also seems pointless."

That almost militantly atheistic view helps some observers explain how science has come into bitter conflict with particular religious groups, especially biblical literalists.

"What accentuates the fault line," said Dr. Ernan McMullin, a Roman Catholic priest who is a former director of the history and philosophy of science program at Notre Dame, is that "the scientists see their science being attacked and they immediately rush to the battlements."

"I think they rather enjoy seeing themselves as a persecuted minority instead of as the dominant force in the culture, which they really are," he said.

The Future

Urgent Goals For Governments

Industry looks to short-term goals and has proven highly adept at using science to take care of itself and consumers. A far more uncertain issue is whether the federal government can successfully address issues of human welfare that lie well beyond the industrial horizon—years, decades and even centuries ahead.

"Science is still the wellspring of new options," Dr. Gibbons said. "How else are we going to face the issues of the 21st century on things like the environment, health, security, food and energy?"

Some experts believe that despite the gnawing doubts today, the world will be ever more inclined to seek scientific answers to those questions in the decades to come. "It will probably accelerate," said Dr. John H. Marburger III, President Bush's science adviser, "because it will become increasingly obvious that we need this steady infusion of results to sustain our ability to cope with all these social problems."

An urgent goal, experts say, is to develop new sources of energy, which will become vitally important as oil becomes increasingly scarce. Another is to better understand the nuances of climate change, for instance, how the sun and ocean affect the atmosphere. Such work is in its infancy. Another is to develop ways of countering the spread of nuclear arms and germ weapons.

The world will also need a new science of cities, to help coordinate planning in areas like waste, water use, congestion, highways, hazard mitigation and pollution control.

"It's going to take a lot of work," said Dr. Grant Heiken, an editor of "Earth Science in the City," a collection of essays just published by the American Geophysical Union in Washington. The number of urban dwellers is expected to grow from three billion now to five billion by 2025.

"I don't know if we'll get a new science," Dr. Heiken said, "but we damn well better."

Dr. Richard E. Smalley, a Rice University professor and Nobel laureate in chemistry, argues that new technologies and conservation can probably solve the world's energy needs. But success, he said, requires a new army of scientists and engineers.

Like others, Dr. Smalley worries about a significant shift in the demographics of American graduate schools in science and engineering. By 1999, according to the latest figures from the National Science Foundation, the number of foreign students in full-time engineering programs had soared so high that it exceeded, for the first time, the steeply declining number of Americans.

"Where the bright kids and the big action are is in Asia," Dr. Smalley said. "That's great for them. It is not what I would hope for our country and our economic well-being or our national security."

Whether the complex challenges of today generate a new era of scientific greatness, several scientists said, may depend on how a deeply conflicted public answers the question of whether science still matters.

In many ways, it all boils down to "a schism between people who have accepted the modern scientific view of the world and the people who are fighting that," said Dr. David Baltimore, the Nobel Prize-winning biologist who is president of the California Institute of Technology.

"Scientists are presenting a much more complicated, much less ethically grounded view of the world, and it's hard for people to take that in," he added.

Some experts warn that if support for science falters and if the American public loses interest in it, such apathy may foster an age in which scientific elites ignore the public weal and global imperatives for their own narrow interests, producing something like a dictatorship of the lab coats.

"For any man to abdicate an interest in science," Jacob Bronowski, the science historian, wrote, "is to walk with open eyes towards slavery."

IN REVIEW

1. List some of the causes of the tension between science and the public.

2. What are some ways that religion and science have clashed in the past? How are these clashes different from today's?

3. How do political leaders use their attitudes about science to shape political agendas?

4. List some of the ways that scientific breakthroughs have benefited the public and some ways that such breakthroughs have caused significant problems. In your opinion, have the benefits outweighed the problems, or vice versa?

5. How much should the government spend to support scientific research? Should the government spend more if competing foreign governments are also spending more?

In the 1,000 years between the fall of the Roman Empire and the European Renaissance, much scientific thought was maintained and advanced by Islamic scholars. Indeed, without these Arabic and Persian scientists, the ideas of the Ancient Greeks might well have perished. In particular, Middle Eastern astronomers developed new instruments for observing the night sky and were responsible for naming many of the stars visible by the unaided human eye. Their observations and scientific theories also laid the foundation for the Copernican revolution and the beginnings of modern astronomy and physics.

How Islam Won, and Lost, the Lead in Science

By Dennis Overbye
The New York Times, **October 30, 2001**

Nasir al-Din al-Tusi was still a young man when the Assassins made him an offer he couldn't refuse.

His hometown had been devastated by Mongol armies, and so, early in the 13th century, al-Tusi, a promising astronomer and philosopher, came to dwell in the legendary fortress city of Alamut in the mountains of northern Persia.

He lived among a heretical and secretive sect of Shiite Muslims, whose members practiced political murder as a tactic and were dubbed hashishinn, legend has it, because of their use of hashish.

Although al-Tusi later said he had been held in Alamut against his will, the library there was renowned for its excellence, and al-Tusi thrived there, publishing works on astronomy, ethics, mathematics and philosophy that marked him as one of the great intellectuals of his age.

But when the armies of Halagu, the grandson of Genghis Khan, massed outside the city in 1256, al-Tusi had little trouble deciding where his loyalties lay. He joined Halagu and accompanied him to Baghdad, which fell in 1258. The grateful Halagu built him an observatory at Maragha, in what is now northwestern Iran.

Al-Tusi's deftness and ideological flexibility in pursuit of the resources to do science paid off. The road to modern astronomy, scholars say, leads through the work that he and his followers performed at Maragha and Alamut in the 13th and 14th centuries. It is a road that winds from Athens to Alexandria, Baghdad, Damascus and Córdoba, through the palaces of caliphs and the basement laboratories of alchemists, and it was traveled not just by astronomy but by all science.

Commanded by the Koran to seek knowledge and read nature for signs of the Creator, and inspired by a treasure trove of ancient Greek learning, Muslims created a society that in the Middle Ages was the scientific center of the world. The Arabic language was synonymous with learning and science for 500 hundred years, a golden age that can count among its credits the precursors to modern universities, algebra, the names of the stars and even the notion of science as an empirical inquiry.

"Nothing in Europe could hold a candle to what was going on in the Islamic world until about 1600," said Dr. Jamil Ragep, a professor of the history of science at the University of Oklahoma.

It was the infusion of this knowledge into Western Europe, historians say, that fueled the Renaissance and the scientific revolution.

"Civilizations don't just clash," said Dr. Abdelhamid Sabra, a retired professor of the history of Arabic science who taught at Harvard. "They can learn from each other. Islam is a good example of that." The intellectual meeting of Arabia and Greece was one of the greatest events in history, he said. "Its scale and consequences are

The observatory at Maragha, built for al-Tusi.

enormous, not just for Islam but for Europe and the world."

But historians say they still know very little about this golden age. Few of the major scientific works from that era have been translated from Arabic, and thousands of manuscripts have never even been read by modern scholars. Dr. Sabra characterizes the history of Islamic science as a field that "hasn't even begun yet."

Islam's rich intellectual history, scholars are at pains and seem saddened and embarrassed to point out, belies the image cast by recent world events. Traditionally, Islam has encouraged science and learning. "There is no conflict between Islam and science," said Dr. Osman Bakar of the Center for Muslim-Christian Understanding at Georgetown.

"Knowledge is part of the creed," added Dr. Farouk El-Baz, a geologist at

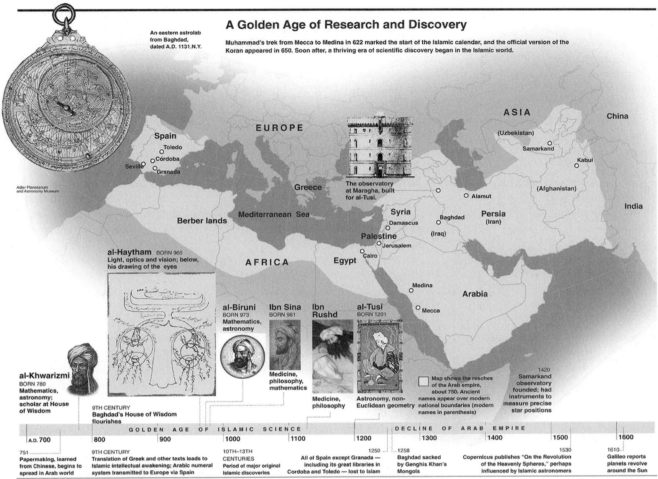

A Golden Age of Research and Discovery

An eastern astrolab from Baghdad, dated A.D. 1131.N.Y.

Muhammad's trek from Mecca to Medina in 622 marked the start of the Islamic calendar, and the official version of the Koran appeared in 650. Soon after, a thriving era of scientific discovery began in the Islamic world.

Adler Planetarium and Astronomy Museum

EUROPE

Spain
Toledo
Córdoba
Seville
Granada

Berber lands

Mediterranean Sea

Greece

The observatory at Maragha, built for al-Tusi.

AFRICA

Egypt

Syria
Damascus

Palestine
Jerusalem
Cairo

Baghdad

(Iraq)

Persia
(Iran)

Alamut

ASIA

(Uzbekistan)
Samarkand

Kabul

(Afghanistan)

China

India

Medina

Arabia

Mecca

al-Haytham BORN 965
Light, optics and vision; below, his drawing of the eyes

al-Khwarizmi
BORN 780
Mathematics, astronomy; scholar at House of Wisdom

al-Biruni
BORN 973
Mathematics, astronomy

Ibn Sina
BORN 981

Medicine, philosophy, mathematics

Ibn Rushd

Medicine, philosophy

al-Tusi
BORN 1201

Astronomy, non-Euclidean geometry

Map shows the reaches of the Arab empire, about 750. Ancient names appear over modern national boundaries (modern names in parenthesis)

1420 Samarkand observatory founded; had instruments to measure precise star positions

9TH CENTURY
Baghdad's House of Wisdom flourishes

	GOLDEN AGE OF ISLAMIC SCIENCE					DECLINE OF ARAB EMPIRE					
A.D. 700	800	900	1000	1100	1200	1300	1400	1500	1600

751
Papermaking, learned from Chinese, begins to spread in Arab world

9TH CENTURY
Translation of Greek and other texts leads to Islamic intellectual awakening; Arabic numeral system transmitted to Europe via Spain

10TH–13TH CENTURIES
Period of major original Islamic discoveries

1250
All of Spain except Granada — including its great libraries in Cordoba and Toledo — lost to Islam

1258
Baghdad sacked by Genghis Khan's Mongols

1530
Copernicus publishes "On the Revolution of the Heavenly Spheres," perhaps influenced by Islamic astronomers

1610
Galileo reports planets revolve around the Sun

Sources: The Columbia History of the World. History of Islamic Science

The New York Times

Boston University, who was science adviser to President Anwar el-Sadat of Egypt. "When you know more, you see more evidence of God."

So the notion that modern Islamic science is now considered "abysmal," as Abdus Salam, the first Muslim to win a Nobel Prize in Physics, once put it, haunts Eastern scholars. "Muslims have a kind of nostalgia for the past, when they could contend that they were the dominant cultivators of science," Dr. Bakar said. The relation between science and religion has generated much debate in the Islamic world, he and other scholars said. Some scientists and historians call for an "Islamic science" informed by spiritual values they say Western science ignores, but others argue that a religious conservatism in the East has dampened the skeptical spirit necessary for good science.

The Golden Age

When Muhammad's armies swept out from the Arabian peninsula in the seventh and eighth centuries, annexing territory from Spain to Persia, they also annexed the works of Plato, Aristotle, Democritus, Pythagoras, Archimedes, Hippocrates and other Greek thinkers.

Hellenistic culture had been spread eastward by the armies of Alexander the Great and by religious minorities, including various Christian sects, according to Dr. David Lindberg, a medieval science historian at the University of Wisconsin.

The largely illiterate Muslim conquerors turned to the local intelligentsia to help them govern, Dr. Lindberg said. In the process, he said, they absorbed Greek learning that had yet to be transmitted to the West in a

serious way, or even translated into Latin. "The West had a thin version of Greek knowledge," Dr. Lindberg said. "The East had it all."

In ninth-century Baghdad the Caliph Abu al-Abbas al-Mamun set up an institute, the House of Wisdom, to translate manuscripts. Among the first works rendered into Arabic was the Alexandrian astronomer Ptolemy's "Great Work," which described a universe in which the Sun, Moon, planets and stars revolved around Earth; Al-Magest, as the work was known to Arabic scholars, became the basis for cosmology for the next 500 years.

Jews, Christians and Muslims all participated in this flowering of science, art, medicine and philosophy, which endured for at least 500 years and spread from Spain to Persia. Its height, historians say, was in the 10th and 11th centuries when three great thinkers strode the East: Abu Ali al-Hasan ibn al-Haytham, also known as Alhazen; Abu Rayham Muhammad al-Biruni; and Abu Ali al-Hussein Ibn Sina, also known as Avicenna.

Al-Haytham, born in Iraq in 965, experimented with light and vision, laying the foundation for modern optics and for the notion that science should be based on experiment as well as on philosophical arguments. "He ranks with Archimedes, Kepler and Newton as a great mathematical scientist," said Dr. Lindberg.

The mathematician, astronomer and geographer al-Biruni, born in what is now part of Uzbekistan in 973, wrote some 146 works totaling 13,000 pages, including a vast sociological and geographical study of India.

Ibn Sina was a physician and philosopher born near Bukhara (now in Uzbekistan) in 981. He compiled a million-word medical encyclopedia, the Canons of Medicine, that was used as a textbook in parts of the West until the 17th century.

Scholars say science found such favor in medieval Islam for several reasons. Part of the allure was mystical; it was another way to experience the unity of creation that was the central message of Islam.

"Anyone who studies anatomy will increase his faith in the omnipotence and oneness of God the Almighty," goes a saying often attributed to Abul-Walid Muhammad Ibn Rushd, also known as Averroes, a 13th-century anatomist and philosopher.

Knocking on Heaven's Door

Another reason is that Islam is one of the few religions in human history in which scientific procedures are necessary for religious ritual, Dr. David King, a historian of science at Johann Wolfgang Goethe University in Frankfurt, pointed out in his book "Astronomy in the Service of Islam," published in 1993. Arabs had always been knowledgeable about the stars and used them to navigate the desert, but Islam raised the stakes for astronomy.

The requirement that Muslims face in the direction of Mecca when they pray, for example, required knowledge of the size and shape of the Earth. The best astronomical minds of the Muslim world tackled the job of producing tables or diagrams by which the qibla, or sacred directions, could be found from any point in the Islamic world. Their efforts rose to a precision far beyond the needs of the peasants who would use them, noted Dr. King.

Astronomers at the Samarkand observatory, which was founded about 1420 by the ruler Ulugh Beg, measured star positions to a fraction of a degree, said Dr. El-Baz.

Islamic astronomy reached its zenith, at least from the Western perspective, in the 13th and 14th centuries, when al-Tusi and his successors pushed against the limits of the Ptolemaic world view that had ruled for a millennium.

According to the philosophers, celestial bodies were supposed to move in circles at uniform speeds. But the beauty of Ptolemy's attempt to explain the very ununiform motions of planets and the Sun as seen from Earth was marred by corrections like orbits within orbits, known as epicycles, and geometrical modifications.

Al-Tusi found a way to restore most of the symmetry to Ptolemy's model by adding pairs of cleverly designed epicycles to each orbit. Following in al-Tusi's footsteps, the 14th-century astronomer Ala al-Din Abul-Hasan ibn al-Shatir had managed to go further and construct a completely symmetrical model.

Copernicus, who overturned the Ptolemaic universe in 1530 by proposing that the planets revolved around the Sun, expressed ideas similar to the Muslim astronomers in his early writings. This has led some historians to suggest that there is a previously unknown link between Copernicus and the Islamic astronomers, even though neither ibn al-Shatir's nor al-Tusi's work is known to have ever been translated into Latin, and therefore was presumably unknown in the West.

Dr. Owen Gingerich, an astronomer and historian of astronomy at Harvard, said he believed that Copernicus could have developed the ideas independently, but wrote in Scientific American that the whole idea of criticizing Ptolemy and reforming his model was part of "the climate of opinion inherited by the Latin West from Islam."

The Decline of the East

Despite their awareness of Ptolemy's flaws, Islamic astronomers were a long ways from throwing out his model: dismissing it would have required a philosophical as well as cosmological revolution. "In some ways it was beginning to happen," said Dr. Ragep of the University of Oklahoma. But the East had no need of heliocentric models of the universe, said Dr. King of Frankfurt. All motion being relative, he said, it was irrelevant for the purposes of Muslim rituals whether the sun went around the Earth or vice versa.

From the 10th to the 13th century Europeans, especially in Spain, were translating Arabic works into Hebrew and Latin "as fast as they could," said Dr. King. The result was a rebirth of learning that ultimately transformed Western civilization.

Why didn't Eastern science go forward as well? "Nobody has answered that question satisfactorily," said Dr. Sabra of Harvard. Pressed, historians offer up a constellation of reasons. Among other things, the Islamic empire began to be whittled away in the 13th century by Crusaders from the West and Mongols from the East.

Christians reconquered Spain and its magnificent libraries in Córdoba and Toledo, full of Arab learning. As a result, Islamic centers of learning began to lose touch with one another and with the West, leading to a gradual erosion in two of the main pillars of science—communication and financial support.

In the West, science was able to pay for itself in new technology like the steam engine and to attract financing from industry, but in the East it remained dependent on the patronage and curiosity of sultans and caliphs. Further, the Ottomans, who took over the Arabic lands in the 16th century, were builders and conquerors, not thinkers, said Dr. El-Baz of Boston University, and support waned. "You cannot expect the science to be excellent while the society is not," he said.

Others argue, however, that Islamic science seems to decline only when viewed through Western, secular eyes. "It's possible to live without an industrial revolution if you have enough camels and food," Dr. King said.

"Why did Muslim science decline?" he said. "That's a very Western question. It flourished for a thousand years—no civilization on Earth has flourished that long in that way."

Islamic Science Wars

Humiliating encounters with Western colonial powers in the 19th century produced a hunger for Western science and technology, or at least the economic and military power they could produce, scholars say. Reformers bent on modernizing Eastern educational systems to include Western science could argue that Muslims would only be reclaiming their own, since the West had inherited science from the Islamic world to begin with.

In some ways these efforts have been very successful. "In particular countries the science syllabus is quite modern," said Dr. Bakar of Georgetown, citing Malaysia, Jordan and Pakistan, in particular. Even in Saudi Arabia, one of the most conservative Muslim states, science classes are conducted in English, Dr. Sabra said.

Nevertheless, science still lags in the Muslim world, according to Dr. Pervez Hoodbhoy, a Pakistani physicist and professor at Quaid-eAzam University in Islamabad, who has written on Islam and science. According to his own informal survey, included in his 1991 book "Islam and Science, Religious Orthodoxy and the Battle for Rationality," Muslims are seriously underrepresented in science, accounting for fewer than 1 percent of the world's scientists while they account for almost a fifth of the world's population. Israel, he reports, has almost twice as many scientists as the Muslim countries put together.

Among other sociological and economic factors, like the lack of a middle class, Dr. Hoodbhoy attributes the malaise of Muslim science to an increasing emphasis over the last millennium on rote learning based on the Koran.

"The notion that all knowledge is in the Great Text is a great disincentive to learning," he said. "It's destructive if we want to create a thinking person, someone who can analyze, question and create." Dr. Bruno Guideroni, a Muslim who is an astrophysicist at the National Center for Scientific Research in Paris, said, "The fundamentalists criticize science simply because it is Western."

Other scholars said the attitude of conservative Muslims to science was not so much hostile as schizophrenic, wanting its benefits but not its world view. "They may use modern technology, but they don't deal with issues of religion and science." said Dr. Bakar.

One response to the invasion of Western science, said the scientists, has been an effort to "Islamicize" science by portraying the Koran as a source of scientific knowledge.

Dr. Hoodbhoy said such groups had criticized the concept of cause and effect. Educational guidelines once issued by the Institute for Policy Studies in Pakistan, for example, included the recommendation that physical effects not be related to causes.

For example, it was not Islamic to say that combining hydrogen and oxygen makes water. "You were supposed to say," Dr. Hoodbhoy recounted, "that when you bring hydrogen and oxygen together then by the will of Allah water was created."

Even Muslims who reject fundamentalism, however, have expressed doubts about the desirability of following the Western style of science, saying that it subverts traditional spiritual values and promotes materialism and alienation.

"No science is created in a vacuum," said Dr. Seyyed Hossein Nasr, a science historian, author, philosopher and professor of Islamic studies at George Washington University, during a speech at the Massachusetts Institute of Technology a few years ago. "Science arose under particular circumstances in the West with certain philosophical presumptions about the nature of reality."

Dr. Muzaffar Iqbal, a chemist and the president and founder of the Center for Islam and Science in Alberta, Canada, explained: "Modern science doesn't claim to address the purpose of life; that is outside the domain. In the Islamic world, purpose is integral, part of that life."

Most working scientists tend to scoff at the notion that science can be divided into ethnic, religious or any other kind of flavor. There is only one universe. The process of asking and answering questions about nature, they say, eventually erases the particular circumstances from which those questions arise.

In his book, Dr. Hoodbhoy recounts how Dr. Salam, Dr. Steven Weinberg, now at the University of Texas, and Dr. Sheldon Glashow at Harvard, shared the Nobel Prize for showing that electromagnetism and the so-called weak nuclear force are different manifestations of a single force.

Dr. Salam and Dr. Weinberg had devised the same contribution to that theory independently, he wrote, despite the fact that Dr. Weinberg is an atheist while Dr. Salam was a Muslim who prayed regularly and quoted from the Koran. Dr. Salam confirmed the account in his introduction to the book, describing himself as "geographically and ideologically remote" from Dr. Weinberg.

"Science is international," said Dr. El-Baz. "There is no such thing as Islamic science. Science is like building a big building, a pyramid. Each person puts up a block. These blocks have never had a religion. It's irrelevant, the color of the guy who put up the block."

IN REVIEW

1. What are some of the ways that Islam promotes science?

2. List the major advances that Islamic scientists and mathematics made between 500 and 1500 A.D.

3. After Muslim conquerors swept through the Middle East, Northern Africa, and Spain, how did their actions and policies encourage intellectual thought?

4. What are some of the factors that may have prevented Middle Eastern intellectuals from making the advances in science that occurred during the European Renaissance?

5. Why is science fundamentally international, and why does science progress despite differences in religions throughout the world?

Isaac Newton, who first understood and described the basic laws of motion, is from a different time and world from University of Nebraska football. But by looking at how players collide on the field and how the football flies through the air, physicists can illustrate some of the fundamental concepts that also describe the motions of the Moon, planets, and stars. The terms *velocity*, *acceleration*, *mass*, *force*, and *momentum* come alive on the football field in ways appreciated by fans and scientists alike. During games, a physics professor at Nebraska uses football as a "hook" to increase the fans' understanding of Newton's laws of motion and fundamental physics.

Crunch! Oof! Well, That's Physics

By Henry Fountain
The New York Times, **November 16, 2004**

It's third and long midway through the second quarter, and Baylor's quarterback arcs a pass 30 yards down the field into Nebraska territory. The ball is thrown in front of the intended receiver, however, and two Nebraska defenders converge on it from opposite directions. Their eyes on the ball and not on each other, they collide at nearly full tilt and the ball pops away.

To the 77,000 fans at Memorial Stadium on this October Saturday, all but a handful dressed in red in tribute to their beloved Cornhuskers, this is just a typical bruising hit, made slightly more interesting, and alarming, because it involves two players on their team. But Dr. Timothy Gay sees it differently.

"Wow, cool, a three-body collision!" Dr. Gay said from his seat in the stands. The forces in this encounter are enormous, but the players don't appear to be injured. Their pads and helmets and the shortness of the collision help protect them, and the third "body"—the ball—absorbs a little bit of the momentum.

Weekdays, Dr. Gay is an experimental atomic physicist at the university who spends most of his time smashing electrons in a basement laboratory, studying the way they scatter as a means of understanding what might go on in the plasma of a fusion reactor or a star.

On fall weekends, when the Huskers play, he makes the short walk across campus to Memorial Stadium, to pursue his avocation—football physics.

Photo by David Weaver for The New York Times

Dr. Timothy Gay watches college football games through the eyes of a physicist.

To watch a football game with Dr. Gay is to view the sport through a different lens, one where talk of fly patterns, blitzes and muffed punts is supplemented by discussions of vector analysis, conservation of momentum and strange forces that can affect the flight of the ball. At Dr. Gay's perch a dozen rows back on the 35-yard line, Isaac Newton is cited as often as Vince Lombardi, and the X's and O's of the game are enhanced by delta-V's and delta-T's.

Back at his lab after the game, he does a quick estimate of the forces involved in that defender-on-defender hit. The players, who weigh about 200 pounds each, are running at about 20 feet per second, and after

they collide they bounce back at perhaps half that speed. It's easy to calculate the acceleration—change in velocity, delta-V, over change in time, delta-T—and force, which by Newton's second law is proportional to mass times acceleration.

The rough result is that the players encounter a force of about 1,800 pounds and an acceleration of 9 g's, or 9 times the force of gravity. Such forces would be bone-breaking and capillary-draining if applied over time, but in the split second of this collision the players can withstand them. The third body helps a little too—with its much smaller mass, the football is sent flying toward the sideline by the momentum imparted to it.

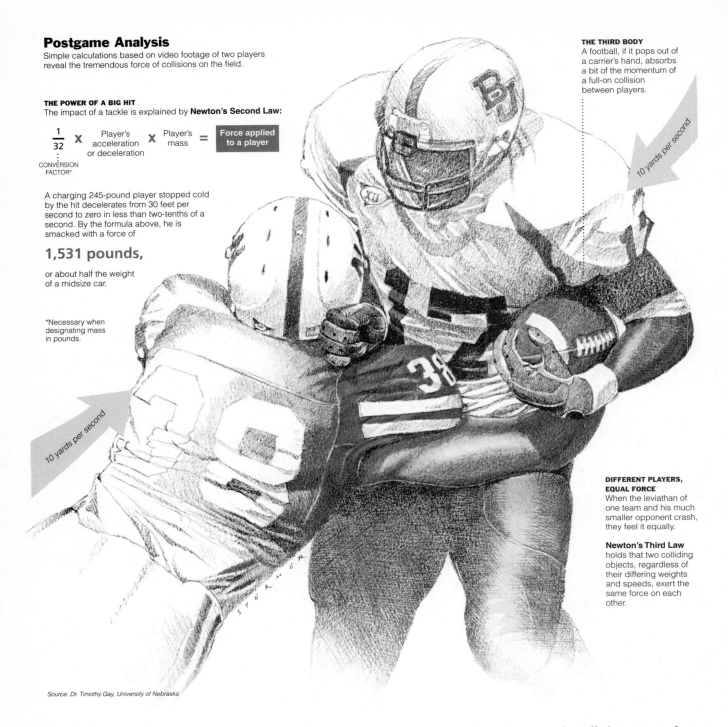

Postgame Analysis

Simple calculations based on video footage of two players reveal the tremendous force of collisions on the field.

THE POWER OF A BIG HIT
The impact of a tackle is explained by **Newton's Second Law:**

$$\frac{1}{32} \times \text{Player's acceleration or deceleration} \times \text{Player's mass} = \boxed{\text{Force applied to a player}}$$

CONVERSION FACTOR*

A charging 245-pound player stopped cold by the hit decelerates from 30 feet per second to zero in less than two-tenths of a second. By the formula above, he is smacked with a force of

1,531 pounds,

or about half the weight of a midsize car.

*Necessary when designating mass in pounds.

THE THIRD BODY
A football, if it pops out of a carrier's hand, absorbs a bit of the momentum of a full-on collision between players.

10 yards per second

10 yards per second

DIFFERENT PLAYERS, EQUAL FORCE
When the leviathan of one team and his much smaller opponent crash, they feel it equally.

Newton's Third Law
holds that two colliding objects, regardless of their differing weights and speeds, exert the same force on each other.

Source: Dr. Timothy Gay, University of Nebraska

Dr. Gay draws a parallel to his work. "The three-body collision—that's the kind of thing I do for a living in the lab," he said. "In atom-molecule collisions, you cannot make a certain chemical reaction go unless you have a third body in there to take up some of the momentum. It's essentially the principle of catalysis."

Dr. Gay is as much a teacher as he is a researcher, and for the past five years has been intent on teaching fans

of football something of the science behind it, first with a series of humorous one-minute videos shown on the scoreboard at Nebraska games and now with a book, "Football Physics: The Science of the Game."

"My connection to football is deep because what I do is collisions," he said. "I'm really interested in what happens if I send an electron in, where it's going to scatter to, how much momentum will it transfer."

"You see that all the time in football," he added. "You see guys colliding. Obviously the physics is a bit different. In football we use Newtonian physics, in atomic collisions we use quantum mechanics."

Using Newtonian physics, he explains later in the game why it was so easy for Nebraska to score on a short goal-line plunge.

"It's just the classic advantage of the momentum of the offense," Dr. Gay said. Because the offense knows when the ball is going to be snapped and the defense doesn't, the offensive line has about two-tenths of a second to build up momentum before the defense can react.

In Newtonian terms, momentum is simply mass times speed. The Nebraska linemen are both large (on the order of 300 pounds each) and speedy (they can run a 40-yard dash in about 5 seconds), so in that two-tenths of a second the line has built up a lot of momentum—something like 10,000 pound-mass-feet per second in the sometimes arcane units of physics.

Dr. Gay put it more plainly: "They've got a head of steam so they can just bowl over the defense, which hasn't started moving yet. It's Newton's first law in action."

Dr. Gay traces his interest in football and physics to his years at prep school in Massachusetts. He wasn't good enough to make the squad, but he worked as team manager. At the same time, he took his first serious physics course. The two interests coalesced.

"Plus, this was football in New England in the fall," Dr. Gay said. "It doesn't get any better than that."

Of course, it has gotten better. With degrees from Caltech (where he was good enough, or rather the team was bad enough, that he played on the offensive line) and the University of Chicago, Dr. Gay has for the past 11 years been at Nebraska, home to one of college football's blue-chip programs and some of its most loyal fans, who have sold out every home game since 1962. Lincoln is the state capital and a college town, but every fall it is swept up in football mania. Even the portable toilets are done up in Cornhusker red.

This year, after decades of a "three yards and a cloud of dust" philosophy, in which a strong running game is crucial, Nebraska has a new coach and a new way of doing things. It's called the West Coast offense, and it relies far more on the passing game, with the receivers running quick, precise routes and the quarterback timing his throws.

In physics terms, the team has gone from relying on mass and force to em-phasizing kinematics, the science of motion and time. The problem is, most of the players were recruited for the old system. So in football terms, the team has gone from good to mediocre—at least by Nebraska's standards.

In the previous week's game, momentum may have been conserved, but pride was not. "Everybody knew we were going to have a tough year," Dr. Gay said. "But we didn't expect to lose, 70-10, to Texas Tech."

This week, though, Nebraska is having an easier time with Baylor, a perennially weak opponent. After a game in which fans pleaded with the team not to throw the football, Nebraska has resorted to more of a ground attack, led initially by running back Cory Ross, who at 5 feet 6 inches and 195 pounds has a low center of mass and is hard to bring down. The Cornhusker line is also throwing its mass around, and Baylor is getting the worst of most collisions. "They're losing the battle of Newton's first law," Dr. Gay said.

That first law—and the second and third, for that matter—are old hat to physicists. "Many of my colleagues say, 'Tim, why are you doing this? This really isn't a very interesting problem, this football physics,'" Dr. Gay said.

"In a narrow sense, they're right," he added. "It's not like we're discovering new physics."

But while it's often described as a collision sport, football is not just about hitting and getting hit. The flight of the ball, for one thing, presents interesting issues. "In this case, I really think there is something new to be learned," he said.

On a basic level, a ball's trajectory has much to do with how tight the spiral is. A wobbly pass presents more surface to the wind, incurring more drag and failing to travel as far. That's one reason, no doubt, why that second quarter Baylor pass fell short of its target. A tight spiral generally requires that the ball be thrown faster, for reasons of torque. "The harder you throw it, the more torque you apply as it leaves your hand, so it spins faster," Dr. Gay said. "That means that it's more stabilized, and you get a tighter spiral."

Dr. Gay was particularly interested in something an old prep school colleague, now a coach for the New England Patriots, told him about punts—how they tend to drift to the right or the left, depending on the direction of spin and whether they "turn over," dipping front-end down after they reach the high point of their arc.

Drawing on work by another physicist, Dr. Marianne Breinig of the University of Tennessee, Dr. Gay described the forces at work. As the spinning ball moves through the air, one side is moving in the same direction as the air moving past it, while the other is moving in the opposite direction. This results in different relative velocities, and "you actually get a pileup of air on one side," he said.

"Because the friction and drag force are bigger, you get turbulence and a force that shoves the ball." This is the Magnus force, which is what makes a spinning baseball curve.

The Magnus effect doesn't create problems on kickoffs and field goal attempts, he said, because the ball is rotating on a different axis, end over end. It might cause the ball to move up or down, but not side to side, so accuracy wouldn't be affected.

As physics concepts go, the Magnus force is fairly ethereal. Back on the field, the situation is much more concrete: Nebraska is thumping Baylor. The Huskers go on to win by 59-27, aided by Ross and another back, Brandon Jackson, who have scored three touchdowns between them. Dr. Gay, objective physicist, is also a subjective, and happy, fan, left to wonder what it is that makes players like Ross and Jackson so good.

Science, he finds, offers only a partial explanation. "As biomechanical machines they're so complex," he said. "You can talk about overall patterns and features and stuff like that, but it's difficult to say why one running back is better than another.

"You know, heart, the will to win. Physics isn't going to touch that."

IN REVIEW

1. How are Newton's laws of motion demonstrated by the football game between Nebraska and Baylor?

2. The collisions among the football players result in transfers of momentum. Explain how momentum is conserved in these collisions.

3. In your opinion, should the physicist use football to illustrate physics to the fans? Explain.

4. Summarize how Newton's laws of motion explain why Nebraska beat Baylor in the football game.

5. How do frictional and drag forces affect the flight of the football? How are Newton's first and third laws demonstrated by the Magnus effect, which affects thrown, but not kicked, footballs during flight?

Before Albert Einstein stole the spotlight, Isaac Newton was the undisputed "King of Physics." Newton's laws of motion and universal gravitation gave scientists the fundamental equations to understand the workings of the solar system, and ultimately enabled engineers to design the spacecraft that enabled humans to land safely on the Moon and robots to land on Venus, Mars, and Saturn's moon Titan. Behind the genius of Newton is a compelling story of a man who forever changed our view of the universe—a story that ultimately was captured in literature, art, poetry, and music. In this way, the influence of Newton advanced not only science, but cultural history as well.

The Man Who Grasped The Heavens' Gravitas

By John Noble Wilford
The New York Times, October 8, 2004

Correction Appended

If Einstein is today's personification of scientific genius, he inherited that exalted role from none other than Isaac Newton, of whom it was said that this was "the greatest and the luckiest of mortals."

In the tribute, credited to the 18th-century French mathematician Joseph-Louis Lagrange, Newton (1642–1727) was deemed the greatest because he discovered the law of universal gravitation and the luckiest because there was only one universe. His brilliance extended not only to the motions of worlds and falling apples, but to an early system of the calculus and a radical new theory of light and color.

Newton's stature as one of the greatest figures in the history of science, and the influence of his ideas on the wider culture for more than two centuries, is the subject of a thoughtful and engaging exhibition at the New York Public Library. The show, "The Newtonian Moment: Science and the Making of Modern Culture," opens today and will run through Feb. 5.

"Newton was immensely curious and obviously immensely talented," said Dr. Mordechai Feingold, a history professor at the California Institute of Technology who is curator of the exhibition, as he conducted a tour of the hundreds of rare books, prints, instruments and other displays of Newtonian science.

A number of Newton's manuscripts from the Cambridge University Library, including a first edition of his most famous work, "Principia," bearing the author's corrections and additions for the next printing, have never before been shown in the United States. Dr. Feingold has also written a companion book to the exhibition, to be published this fall by Oxford University Press.

The Newtonian Moment began in the plague years of 1665-66 when the young scholar first formulated many of his revolutionary ideas of mathematics, optics and mechanics. He had already absorbed the essential science of Descartes, Galileo and Robert Hooke. As Dr. Feingold observed, much of Newton's genius consisted of "his remarkable ability to simultaneously consume and transform any knowledge he acquired."

Newton himself attributed at least some of his success to the fact that he had stood "on the shoulders of giants."

Newton came to prominence in 1671 when a small telescope he designed and built won him election to the Royal Society. Then Edmond Halley, he of comet fame, went to Cambridge to ask Newton's advice on the shape of orbits traveled by planets. Ellipses, Newton replied forthwith. He had already figured it out.

Impressed, Halley persuaded a reluctant Newton to write what became his magnum opus, "Principia,"

published in 1687. His other great work, "Opticks," followed in 1704, describing how a beam of light, when passed through a prism, dispersed into the many colors of the visual spectrum.

The exhibition dwells at length on the international response to Newton's science. Germans contended that Leibniz was the true inventor of calculus, and his principles are indeed the basis for today's calculus. The French stoutly defended their Descartes, whose universe was full of matter whirling in vortices. The ascendant Dutch universities served as arbiters in Newton's ultimate triumph.

The French, to their credit, eventually organized expeditions to Lapland and Peru that confirmed a prediction of Newtonian gravity: the Earth is an oblate spheroid, flattened at the poles and bulging at the equator.

The complex character of Newton is also explored in the exhibition. He was a pious man who dabbled in chronologies of biblical history, though he apparently held unorthodox views on Christian doctrine like the Trinity. He also devoted much effort to alchemy, the practice of trying to turn base metals to gold, which makes Newton a central character in a new novel, "The System of the World," by Neal Stephenson (William Morrow). Dr. Feingold suggests that

Newton's alchemy was probably the contemporary name for what today would be standard chemistry.

In many ways, the exhibition's most fascinating displays highlight the dissemination of Newtonian ideas beyond the narrow world of science, which contributed to their transforming influence on the entire culture and the idolization of genius himself. He came to be considered, in Dr. Feingold's phrase, "the acme of human possibility."

"Science was a minor partner in culture in 1600," Dr. Feingold said on the tour. "By 1800, it was the major part of culture."

For more than a century after his death in 1727, Newton was immortalized by poets, sculptors, painters and other worthies whose works are on display. Alexander Pope's familiar couplet eulogized Newton: "Nature and Nature's Laws lay hid at Night. / God said, Let Newton be! and All was Light." He is similarly memorialized in prints and reproductions by William Hogarth and numerous other artists.

Newton's bust had a prominent place in aristocratic gardens, and often appeared in the background of portraits of scientists and men of letters, notably one of Benjamin Franklin. Even Newton's critics recognized his towering intellect. Under a color print by William Blake, showing Newton seated on a rock in the "sea of time and space," the description label notes that for Blake, "Newton is the misguided genius whose mechanical universe left no room for the imagination or for God, but who could ultimately find a prominent place in heaven."

Another Newton legacy, amply illustrated in the displays, was the popularization of science. The trend became especially widespread in the 1740's with the fascination about the nature of electricity. A print of one electrical experiment anticipated Rube Goldberg. A boy, suspended by silk cords, is shown drawing an electric charge from a generator. He transmits the charge to the girl standing on dried pitch, allowing her to attract and repel chaff.

Several arresting displays show the actual devices or drawings of mechanical models of the solar system that were popular in the post-Newton period. Such an elaborate device, known as the orrery, illustrated the relative size and motion of heavenly bodies in accordance with Newton's laws of gravity.

Writers like Francesco Algarotti also facilitated the passage of Newton's ideas to the general public. He translated the recondite language of Newton's "Principia" into a dialogue format interspersed with amusing digressions. The book was written specifically to appeal to women. Other books of the time, like the Tom Telescope series, introduced Newton to children.

The most famous popularizer of Newtonian science was Voltaire, who in 1738 published the book "Elémens de la Philosophie de Neuton." The book's allegorical frontispiece typifies the virtual deification of Newton. Voltaire is seen writing at his desk. Over him a shaft of light from heaven, the light of truth, passes through Newton to the author's lover and collaborator, Madame du Châtelet. She reflects the light onto the inspired Voltaire.

The acknowledgment of Madame du Châtelet's role ("Minerva dictated and I wrote," Voltaire said) was well deserved. Voltaire's muse happened to be a versatile natural philospher who translated "Principia" into French, accompanied by an extensive commentary, and made no attempt to conceal her talents. She once told Frederick the Great of Prussia, "It may be that there are metaphysicians and philosophers whose learning is greater than mine, although I have not met them."

Voltaire's book was a great success. He commanded a wide audience, Dr. Feingold said, because "he was neither a mathematician nor a physicist, but a literary giant aloof from the academic disputes over Newtonian ideas." In other words, the historian added, Voltaire's amateurism in science "was a source of his contemporary appeal, demonstrating for the first time the accessibility of Newton's ideas to nonspecialists."

So a journalist who makes his living reporting and explaining the labors of scientists comes away from the exhibition both elevated and humbled. He can count Voltaire among his most celebrated forerunners. Even science journalists stand on the shoulders of giants.

He Saw the Light

"The Newtonian Moment: Science and the Making of Modern Culture," is on view through Feb. 5 in the Gottesman Exhibition Hall in the Humanities and Social Sciences Library of the New York Public Library, Fifth Avenue and 42nd Street. Hours: Tuesdays and Wednesdays, 11 a.m. to 7:30 p.m.; Thursdays through Saturdays, 10 a.m. to 6 p.m. Admission is free. Information: (212) 869-8089

Two lectures accompany the exhibition; tickets to each are $10. "Man of the Moment: How Newton Moved Mathematics to the Top of the Scientific Agenda," by Lisa Jardine, a professor of Renaissance Studies at Queen Mary, University of London will be given Nov. 30 at 6:30 p.m. "The Scientist as Scholar: Newton as Historian," by Anthony Grafton, a professor of history at Princeton University, will be on Jan. 5 at 6:30 p.m.

Correction: October 18, 2004, Monday. An article in Weekend on Oct. 8 about the Isaac Newton exhibition at the New York Public Library included the library's misstatement of part of a couplet by Alexander Pope. His eulogy of Newton read "Nature and Nature's Laws lay hid in Night. / God said, Let Newton be! and all was Light." (Pope did not write "hid at Night.")

IN REVIEW

1. Newton attributes some of his genius to the earlier work of others. How is this statement a reflection of the processes of science?

2. What life events described in the article are attributable to some of Newton's scientific breakthroughs? In your opinion, would he have made his important discoveries about gravitation, force, and motion without these life events? Explain.

3. Newton was immortalized by "poets, sculptors, and painters" after his death. Explain why someone would want to use art to immortalize the works of a scientist.

4. Newton was responsible, in large part, for the popularization of science in the 1700s and 1800s. How did artists using Newton's scientific theories help in this popularization?

5. Einstein eclipsed Newton in popularity in the early 1900s. What aspects of Einstein's work might have made him more popular? What societal and technological changes in the early 1900s might also have contributed to Einstein's popularity?

Gravity is the force of attraction between masses. Earth, with its tremendous mass, creates a force of gravity that is responsible both for keeping us safely on the planet's surface and for keeping the Moon in orbit. However, the Earth's relatively large gravitational force makes it difficult, and therefore costly, to launch spacecraft into orbit and beyond. Indeed, some attribute the current delay in human exploration of the solar system to the large costs associated with escaping the Earth's gravity. This idea is clearly illustrated in a quote by science fiction author Robert Heinlein: "Once you get to Earth orbit, you're halfway to anywhere in the solar system." A solution to the high cost of getting objects into Earth's orbit may be the space elevator, which has been promoted by author Arthur C. Clarke for many years and may now be possible because of the creation of certain high-tech materials.

Not Science Fiction: An Elevator to Space

By Kenneth Chang
The New York Times, **September 23, 2003**

With advances toward ultrastrong fibers, the concept of building an elevator 60,000 miles high to carry cargo into space is moving from the realm of science fiction to the fringes of reality.

This month, the Los Alamos National Laboratory was a sponsor of a conference to ponder the concept. Yet, the keynote address was by a titan of science fiction, Arthur C. Clarke, speaking via satellite from his home in Sri Lanka. "I'm happy that people are taking it more and more seriously," said Mr. Clarke, whose novel "The Fountains of Paradise" (1978) revolved around such a space elevator.

The discovery in 1991 of nanotubes, cylindrical molecules of carbon with many times the strength of steel, turned the idea from a fantastical impossibility to an intriguing possibility that could be realized in as little as a decade or two.

Proponents say the economic and technological advantages of a space elevator over rockets make it inevitable. They predict it will lower the cost of putting a satellite into space from $10,000 a pound to $100.

"As soon as we can build it, we should build it," said Dr. Bryan E. Laubscher, a scientist at Los Alamos who organized the conference. Just as the transcontinental railroad opened the West in the late 1800's, "I feel the space elevator is going to be such a paradigm shift in space access," Dr. Laubscher said.

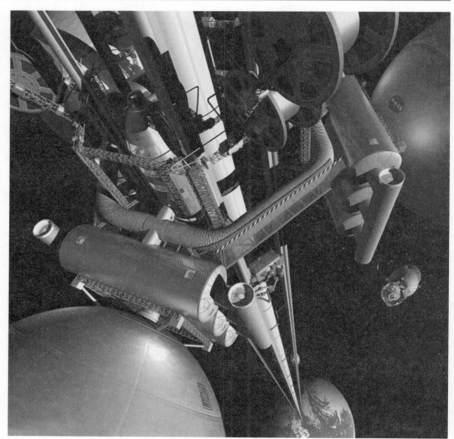

Photo by Flight Projects Directorate

Scientists and engineers are working on the concept of an elevator to space.

Easier economical access to space might also make practical other grandiose projects like solar power satellites that could collect sunlight and beam energy down to Earth.

The conference, a three-day session here, drew 60 people, a mix of scientists and engineers who are working on the concept, space enthusiasts who wanted to hear more and dilettantes from nearby Los Alamos laboratory attracted by curiosity.

"The first thought is, Is this really going to work?" said Dr. Steven E.

Patamia, a researcher at Los Alamos, who was enlisted into performing space elevator calculations a week before the conference. "When you get into it, it begins to make sense. There are a good number of technical issues. They are probably all 'overcomeable.'"

The original idea of a space elevator is more than a century old. In 1895, Konstantin E. Tsiolkovsky, a Russian visionary who devised workable ideas for rocket propulsion and space travel decades before others, proposed a tower thousands of miles high attached to a "celestial castle" in orbit around Earth, with the centrifugal force of the orbiting castle holding up the tower. (Imagine swinging a rope with a rock tied to the end of it.)

But the idea was fundamentally impossible to build. Steel, then the strongest material known, was too heavy and not strong enough to support that weight.

Other scientists periodically revisited and reinvented Tsiolkovsky's idea, inspiring science fiction writers like Mr. Clarke.

Nanotubes spurred NASA to take a more serious look in 1999. A team of scientists envisioned huge cables of nanotubes and magnetically levitated cars traveling up and down. The structure would be so large that it would require grabbing an asteroid and dragging it into Earth orbit to act as the counterweight for holding up the elevator.

To avoid weather, especially lightning, the NASA scientists envisioned the base station as a tower at least 10 miles high.

"We came out of that workshop saying the space elevator is 50 years away," said David V. Smitherman of the Marshall Space Flight Center, who led the study.

Around that time, Dr. Bradley C. Edwards, who was then a scientist at Los Alamos, read an even more pessimistic assessment, that a space elevator would not be built for at least 300 years.

"But there was no information why it couldn't be built," Dr. Edwards said, and he took that as a challenge.

Dr. Edwards simplified the NASA idea to what he calls "the Wright brothers' version," a single ribbon

Going Up

How a 60,000-mile elevator for space cargo could be constructed.

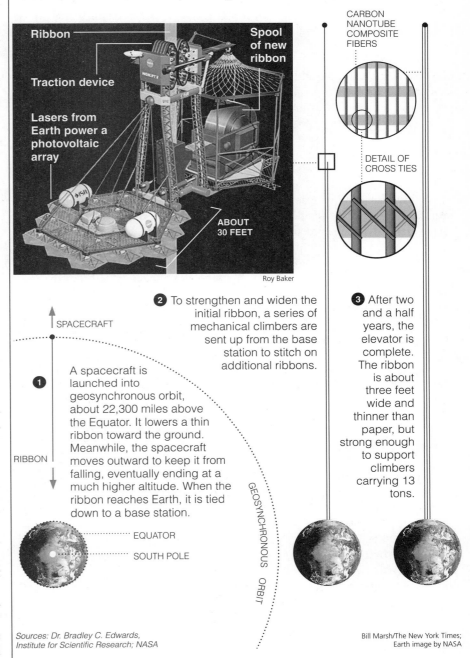

Ribbon
Traction device
Lasers from Earth power a photovoltaic array
Spool of new ribbon
ABOUT 30 FEET
Roy Baker

CARBON NANOTUBE COMPOSITE FIBERS
DETAIL OF CROSS TIES

SPACECRAFT

1 A spacecraft is launched into geosynchronous orbit, about 22,300 miles above the Equator. It lowers a thin ribbon toward the ground. Meanwhile, the spacecraft moves outward to keep it from falling, eventually ending at a much higher altitude. When the ribbon reaches Earth, it is tied down to a base station.

RIBBON

EQUATOR
SOUTH POLE

GEOSYNCHRONOUS ORBIT

2 To strengthen and widen the initial ribbon, a series of mechanical climbers are sent up from the base station to stitch on additional ribbons.

3 After two and a half years, the elevator is complete. The ribbon is about three feet wide and thinner than paper, but strong enough to support climbers carrying 13 tons.

Sources: Dr. Bradley C. Edwards, Institute for Scientific Research; NASA

Bill Marsh/The New York Times; Earth image by NASA

about three feet wide and thinner than a piece of paper, stretching 60,000 miles from Earth's surface.

He sent a proposal to the NASA Institute for Advanced Concepts, which provided him $570,000 to flesh out the ideas. His results are described in a book simply titled "The Space Elevator" (Spageo, 2002).

Instead of using magnetic levitation, the apparatus would lift up to

13 tons of cargo by pulling itself upward with a couple of tanklike treads that squeezed tightly onto the ribbon. Up to eight would ascend the ribbon at any one time, powered by lasers on the ground shining on the solar panels on the rising platforms.

It would take about a week for one to reach geosynchronous orbit, 22,300 miles up, where a satellite circles the Earth in exactly one day, continuously hovering over the same spot on the Earth's surface.

The first elevator would go up only. At the top, the platform would simply be added to the counterweight or be discarded into space.

All the necessary underlying technology exists, Dr. Edwards said, except the material for the ribbon. (The longest nanotube to date is just a few feet long.) But he said he expected that scientists would develop a strong enough nanotube-polymer composite in a few years.

"There's a clear path to building this," said Dr. Edwards, now director of research at the Institute for Scientific Research, an independent organization in Fairmont, W.Va. The institute sponsored the conference with Los Alamos.

Dr. Edwards estimates the cost to build the first elevator at $6.2 billion, although with the uncertainties in forecasting a decade or two of research and development, "doubling this is probably a good first cut."

"It's gotten to the point," he added, "where we can say it's closer to $6 billion than $600 billion."

Building subsequent elevators would be cheaper, $2 billion each, because the first elevator could lift materials.

By comparison, the estimated cost of building and operating the International Space Station is widely expected to exceed $100 billion.

Dr. Edwards says he has reasonable solutions for other concerns. The elevator base could be a movable ocean platform in the eastern equatorial Pacific, hundreds of miles from commercial airline routes and easy to defend from terrorist attacks. Hurricanes never cross the Equator, and lightning is sparse in that region. By moving the base station, the elevator operators could drag the apparatus around low-orbit space debris. An aluminum coating could be added to parts of the ribbon to combat decay from the reaction of oxygen with carbon atoms in the nanotubes.

More futuristically, Dr. Edwards imagines that additional elevators can be built on the Moon or Mars, vastly simplifying and speeding spaceflight through the solar system.

"What a wonderful idea if you could ever make it work," said Gentry Lee, the chief engineer of planetary flight systems at NASA's Jet Propulsion Laboratory in Pasadena, Calif., who pushed for financing Dr. Edwards's initial studies. "It is plausible. It is not implausible. I think that the idea has so much promise a couple of million dollars a year aimed at the enabling technologies is not too much to ask."

At the conference, scientists presented calculations that examined details. Vibrations in the elevator ribbon, which would act like an extremely long plucked guitar string, appeared manageable.

Dr. Anders Jorgensen of Los Alamos raised concerns that as the ribbon swung around through the Earth's magnetic field, it would create strong electric currents. Because of the elevator's relatively slow pace, a larger problem could be that any human passengers would receive dangerous doses of radiation as they passed through pockets of high-energy particles trapped in Earth's magnetic field.

The first space elevator would be built to carry only cargo, not people, Dr. Edwards said. The dangers could be reduced on subsequent elevators by speeding them up or providing shielding through magnetic fields.

The logistics of construction "appear to be an interesting problem, as well," said Carey R. Butler, a program manager at the Institute for Scientific Research. The idea is to launch a spacecraft with the initial spools of ribbon into geosynchronous orbit. As the ribbon unspools and falls to the ground, the spacecraft moves higher to keep the center of mass, or balance point, at the same geosynchronous height. By the time the ribbon reaches the ground, the spacecraft has reached an altitude of 50,000 miles.

After the ribbon is attached to the base station, the spacecraft unfurls another 10,000 miles of ribbon. Then a series of about 230 mechanical construction platforms will ascend to stitch on additional ribbon.

While technologically feasible, years of engineering will still be needed. "There's a lot to be done, obviously," Dr. Edwards said.

Nonetheless, Mr. Clarke, who came up with the idea of using satellites in geosynchronous orbit for communications long before any were launched, thought that he might live to see this science fiction idea come true, too.

"I'm 86 now," Mr. Clarke said. "So in 20 years' time, I'll only be 106. So maybe I will see it."

IN REVIEW

1. Describe the basic principles behind the space elevator.

2. How have advances in materials made space elevators possible? What other technological advances are necessary for space elevators to be built?

3. What role have science fiction writers played in promoting the idea of the space elevator? List some other examples of how science fiction helped shape technologies that ultimately became reality.

4. In your opinion, is the cost of building a space elevator justifiable? Explain.

5. Using your knowledge of Newton's laws of motion and universal gravitation, why would a space elevator be a realistic technology for reaching the Moon or Mars?

We generally say that matter can be found in one of three distinct phases: solid, liquid, and gas. While this is a reasonable approximation, reality is more complex. For example, a gas behaves differently if its atoms are ionized, in which case we call it a plasma. Solid phases can also vary, and water (H_2O) ice is an important example, coming in many different known phases.

Moreover, even ordinary water ice is rather unusual: Most substances are more dense in solid than liquid form, but water ice is less dense than liquid water. That is why ice floats—and one of the reasons why water is biologically important. If ice sank, lakes would fully freeze in winter. Because ice floats, the ice forms first on the surface, and this surface ice helps insulate the lake so that the depth remains liquid and fish and other organisms can survive.

This article discusses some of the mysteries of water ice, discussing both everyday questions such as why ice skating is possible and astronomical questions such as the type of ice that might be found on the moons of jovian planets.

Explaining Ice: The Answers Are Slippery

By Kenneth Chang
The New York Times, **February 21, 2006**

Here is one question that probably won't cross the minds of Sasha Cohen, Irina Slutskaya and the other Olympic women figure skaters today, even if they fall: Why is ice slippery?

But maybe it should. After all, ice is a solid, and trying to glide on thin metal blades across the surfaces of most solids — concrete, wood, glass, to name a few — results in ear-piercing sounds and ungraceful stumbles. Though the question may seem to be a simple one, physicists are still searching for a simple answer.

The explanation once commonly dispensed in textbooks turns out to be wrong. And slipperiness is just one of the unanswered puzzles about ice. Besides the everyday ice that you slip on, there are about a dozen other forms, some of which experts suspect exist in the hot interior of Earth or on the surface of Pluto. Scientists expect to discover still more variations in the coming years.

Ice, said Robert M. Rosenberg, an emeritus professor of chemistry at Lawrence University in Appleton, Wis., and a visiting scholar at Northwestern University, "is a very mysterious solid."

Dr. Rosenberg wrote an article looking at the slipperiness of ice in the December issue of Physics Today, because

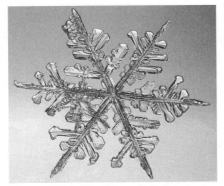

Photo by Edward Kinsman/Photo Researchers Inc.; Illustrations by Al Granberg

In everyday ice, the molecules of the water — H_2O — line up in a hexagonal pattern. That is why snowflakes all have six-sided patterns.

he kept coming across the wrong explanation for it, one that dates back more than a century.

This explanation takes advantage of an unusual property of water: the solid form, ice, is less dense than the liquid form. That is why ice floats on water, while a cube of frozen alcohol — which has a freezing temperature of minus 173 degrees Fahrenheit — would plummet to the bottom of a glass of liquid alcohol. The lower density of ice also means that the melting temperature of ice can be lowered below the usual 32 degrees by squeezing on it.

According to the frequently cited — if incorrect — explanation of why ice is slippery under an ice skate, the pressure exerted along the blade lowers the melting temperature of the top layer of ice, the ice melts and the blade glides on a thin layer of water that refreezes to ice as soon as the blade passes.

"People will still say that when you ask them," Dr. Rosenberg said. "Textbooks are full of it."

But the explanation fails, he said, because the pressure-melting effect is small. A 150-pound person standing on ice wearing a pair of ice skates exerts a pressure of only 50 pounds per square inch on the ice. (A typical blade edge, which is not razor sharp, is about one-eighth of an inch wide and about 12 inches long, yielding a surface area of 1.5 square inches each or 3 square inches for two blades.) That amount of pressure lowers the melting temperature only a small amount, from 32 degrees to 31.97 degrees. Yet ice skaters can easily slip and fall at temperatures much colder.

The pressure-melting explanation also fails to explain why someone wearing flat-bottom shoes, with a much greater surface area that exerts even less pressure on the ice, can also slip on ice.

Don't Fall

What makes Olympic ice skaters slide across the ice? Physicists still disagree over the answer to this seemingly simple question.

PRESSURE MELTING

Theory Applying pressure to ice causes it to melt at a slightly lower temperature. The skater slips on the thin layer of water created by his weight.

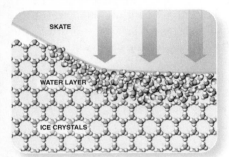

Finding This melting effect occurs. But the change is too small to be the primary reason ice is slippery.

FRICTIONAL HEATING

Theory The fast-moving blade creates friction on the ice, generating heat to melt a thin layer of water under the skate.

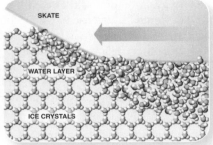

Finding The theory is correct. But it does not explain why a person standing still on ice can also slip.

INTRINSIC SLIPPERY LAYER

Theory A liquid-like film exists on the surface of ice. Chains of water molecules abut the air and are unable to form solid ice crystals.

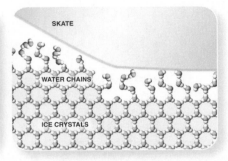

Finding Skaters slip on these molecular chains, which vibrate like water molecules.

Sources: Robert Rosenberg, Physics Today

The New York Times; illustrations by Al Granberg

Two alternative explanations have arisen to take the pressure argument's place. One, now more widely accepted, invokes friction: the rubbing of a skate blade or a shoe bottom over ice, according to this view, heats the ice and melts it, creating a slippery layer.

The other, which emerged a decade ago, rests on the idea that perhaps the surface of ice is simply slippery. This argument holds that water molecules at the ice surface vibrate more, because there are no molecules above them to help hold them in place, and they thus remain an unfrozen liquid even at temperatures far below freezing.

Scientists continue to debate whether friction or the liquid layer plays the more important role. Dr. Rosenberg, asked his opinion, chose a indecisive answer: "I say there are two major reasons."

The notion that ice has an intrinsic liquid layer is not a new concept. It was first proposed by the physicist Michael Faraday in 1850 after a simple experiment: he pressed two cubes of ice against each other, and they fused together. Faraday argued that the liquid layers froze solid when they were no longer at the surface. Because the layer is so thin, however, it was hard for scientists to see.

In 1996, Gabor A. Somorjai, a scientist at Lawrence Berkeley Laboratory, bombarded the surface of ice with electrons and watched how they bounced off, producing a pattern that looked at least partially liquid at temperatures down to minus 235 degrees. A couple of years later, a team of German scientists bounced helium atoms off ice and found results that corroborated the Lawrence Berkeley findings.

"The water layer is absolutely intrinsic to ice," Dr. Somorjai said.

The findings, he said, fit with a simple observation that suggests friction cannot be the one and only explanation of slipperiness. When a person stands on ice, he added, no heat is generated through friction, and yet "ice is still slippery."

But a colleague of Dr. Somorjai's at Lawrence Berkeley, Miquel Salmeron, while he does not dispute Dr. Somorjai's experiment, does dispute the importance of the intrinsic liquid layer to slipperiness.

In 2002, Dr. Salmeron and colleagues performed an experiment. They dragged the tip of an atomic force microscope, resembling a tiny phonograph needle, across the surface of ice.

"We found the friction of ice to be very high," Dr. Salmeron said. That is, ice is not really that slippery, after all.

Dr. Salmeron said that this finding indicates that while the top layer of ice may be liquid, it is too thin to contribute much to slipperiness except near the melting temperature. In his view, friction is the primary reason ice is slippery. (The microscope tip was so small that its friction melted only a tiny bit of water, which immediately refroze and therefore did not provide the usual lubrication, he said.)

Dr. Salmeron concedes, however, that he cannot definitively prove that his view is the correct one.

"It's amazing," he said. "We're in 2006, and we're still talking about this thing."

Ice formed by water behaves even more strangely at lower temperatures and higher pressures.

Water — H$_2$O — seems to be a simple molecule: two hydrogen atoms connected to a central oxygen atom in a V-shape. In everyday ice, which scientists call Ice Ih, the water molecules line up in a hexagonal pattern; this is why snowflakes all have six-sided patterns. (The "h" stands for hexagonal. A variation called Ice Ic, found in ice crystals floating high up in the atmosphere, forms cubic crystals.)

The crystal structure of the ice is fairly loose — the reason that Ice Ih is less dense than liquid water — and the bonds that the hydrogen atoms form between water molecules, called hydrogen bonds, are weaker than most atomic bonds.

At higher pressures, the usual hexagonal structure breaks down, and the bonds rearrange themselves in more compact, denser crystal structures, neatly labeled with Roman numerals: Ice II, Ice III, Ice IV and so on. Scientists have also discovered several forms of ice in which the water molecules are arranged randomly, as in glass.

At a pressure of about 30,000 pounds per square inch, Ice Ih turns into a different type of crystalline ice, Ice II. Ice II does not occur naturally on Earth. Even at the bottom of the thickest portions of the Antarctic ice cap, the weight of three miles of ice pushes down at only one-quarter of the pressure necessary to make Ice II. But planetary scientists expect that Ice II, and possibly some other variations, like Ice VI, exist inside icier bodies in the outer solar system, like the Jupiter moons Ganymede and Callisto.

With pressure high enough, the temperature need not even be cold for ice to form. Several Februaries ago, Alexandra Navrotsky, a professor of chemistry, materials science and geology at the University of California, Davis, was visiting Northwestern. She was sitting in office of Craig R. Bina, a geophysicist, and looking out over frozen Lake Michigan. "Ice might have been on our minds," she recalled.

The scientists started considering what happens to tectonic plates after they are pushed back down into Earth's interior. At about 100 miles down, the temperature of these descending plates is 300 to 400 degrees — well above the boiling point of water at the surface — but cool compared with that of surrounding rocks. The pressure of 700,000 pounds per square inch at this depth, Dr. Bina and Dr. Navrotsky calculated, could be great enough to transform any water that was there into a solid phase known as Ice VII.

No one knows whether ice can be found inside Earth, because no one has yet figured out a way to look 100 miles underground. Just as salt melts ice at the surface, other molecules mixing with the water could impede the freezing that Dr. Bina and Dr. Navrotsky have predicted.

Ice also changes form with dropping temperatures. In hexagonal ice, the usual form, the oxygen atoms are fixed in position, but the hydrogen bonds between water molecules are continually breaking and reattaching, tens of thousands of times a second.

At temperatures cold enough — below minus 330 degrees — the hydrogen bonds freeze as well, and normal ice starts changing into Ice XI.

William B. McKinnon, a professor of earth and planetary sciences at Washington University in St. Louis, said that astronomers were probably already looking at Ice XI on the surface of Pluto and on the moons of Neptune and Uranus. But instruments currently are not sensitive enough to distinguish the slight differences among the ices.

The most recently discovered form of ice, Ice XII, was found just a decade ago, although hints of it had been seen years earlier. John L. Finney of University College London, one of the discoverers of Ice XII, said that trying to understand all the different forms of ice was important for an understanding of how the water molecule works, and that was important in understanding how water interacts with all the biological molecules in living organisms.

"It gives you a very stringent test for our understanding of the water molecule itself," he said.

And could there be an Ice XIII?

"Yes," Dr. Finney said. "Call me in a month."

But scientists have given no word on whether any of these other types of ice are slippery enough to land a triple axel.

IN REVIEW

1. What is the traditional (but wrong) explanation for why ice skating is possible, and why does it fail?

2. How do different phases of ice differ from one another?

3. Could there be any form of water ice deep inside Earth, despite the high temperature? Explain.

4. Why do scientists speculate that Ice II and other phases of water ice might be found inside some of Jupiter's moons?

5. What does the fact that scientists still debate something as seemingly simple as the nature of ice tell you about how science works? Does this fact give you more or less confidence in the methods of science? Defend your opinion.

Telescopes are the primary tools of astronomers. As discussed in your textbook, telescope technology has been advancing rapidly, providing the impetus for many new, important discoveries. This article discusses efforts to build even larger mirrors for the next generation of telescopes.

Mirror, Mirror

By Dennis Overbye
The New York Times, **August 30, 2005**

TUCSON—In the cavernous bowels of the University of Arizona's football stadium, Roger Angel's mirror furnace was spinning like a captured flying saucer at a stately five revolutions per minute.

It was a contrivance that Monty Python or Doc Ock might have designed — 30 feet across and 10 feet high, carapaced with red boxes, steel beams, black cables, flashing lights and metal air ducts snaking from its body like octopus arms.

An orange glow, from 18 tons of molten glass heated to 2,100 degrees Fahrenheit, was peeking through openings around the ducts as they flashed by.

That glass was on its way to being part of the heart of what could be the largest telescope in the world 10 years from now. And so, nearby, several dozen sweltering astronomers and other dignitaries were roaming catwalks, wandering among giant mirrors and mirror polishing machines and swigging bottled water while they kept a weather eye on monitors showing what was going on inside the furnace.

One camera was focused on a set of marks on the furnace wall — not unlike the ones on a child's closet door — used to gauge the level of the molten glass inside. The level had been falling in the last day as the temperature ramped up and chunks of glass the size of cobblestones softened and began to flow down into narrow channels forming a honeycomb pattern.

The glass would stop falling when it had completely filled the honeycomb structure. Meanwhile, centrifugal force would have whipped the overflow into a perfect parabola 28 feet across — the

Photo by Lori Stiles/University of Arizona

A spin-casting furnace casts the mold to build the mirror for the Giant Magellan Telescope.

desired shape for sweeping up starlight dispersed into foggy invisibility over billions of light-years and compressing it into crisp bright dots astronomers could read like a newspaper to learn what was happening around a distant sun or when the universe was born.

That was the moment the real work could begin.

"This project is very gutsy," said Dr. Angel, a slender, gray-haired astronomer who runs the Steward Observatory Mirror Laboratory. He has been building mirrors and populating mountaintops with telescopes this way for 20 years, but nobody has ever built something like this.

If everything works out, the mirror now forming in Dr. Angel's saucerlike

furnace will be only the first of seven making up a giant telescope with the light-gathering power of a mirror 70 feet across. The Giant Magellan, as it is called, would be twice the size of anything now operating on Earth or in space, and four times as powerful. But there are many challenges. To blend their light at a common focus, Dr. Angel explained, all seven mirrors will have to be part of the same giant parabola. That means that all of them except the central mirror must have an unusual "wickedly curved" asymmetrical shape.

And there is the cost. The Giant Magellan will cost half a billion dollars — money that its collaborators, a consortium of eight institutions, does not yet have.

Ceramic fiber cores are installed to complete the mold for the mirror.

To show that they can make such a mirror, and perhaps shake loose some of that half billion, the collaborators — which include the Carnegie Institution of Washington; Harvard; the Massachusetts Institute of Technology; the Smithsonian Astrophysical Observatory; the Universities of Arizona, Michigan and Texas; and Texas A&M — announced this year that they would go ahead and make one, at a cost of some $17 million, and they invited everyone to watch.

"Everybody in collaboration believes we need to test this technology," said Wendy Freedman, director of the Carnegie Observatories and chairwoman of the Giant Magellan board, adding that if the test fails the project will not proceed.

Robert Kirshner of Harvard said, "It's kind of brave to get started before you know you're going to finish."

Dr. Freedman added that they had to start making mirrors now, money or not, to meet their goal of beginning limited operations in Chile in 2013 and finish in 2016.

Making that date will allow them to overlap with the National Aeronautics and Space Administration's James Webb Space Telescope, scheduled for a 2011 launching and keep pace with their rivals, a consortium including the California Institute of Technology, the University of California and the Canadian Astronomical Association that wants to build a telescope 100 feet in diameter, using a radically different technology.

The result in late July was a weekend rendezvous in the desert, part fund-raising party, part seminar on telescope making and part family reunion. Many participants had worked together on other projects, like Magellan, the new telescope's namesake, which consists of twin 21-foot-diameter telescopes at Las Campanas, a Carnegie observatory in Chile, and the Large Binocular Telescope being built on Mount Graham in Arizona.

"Carnegie is returning to its roots," said Richard Meserve, the Carnegie Institution's president in a talk, recalling that it was Carnegie telescopes that Edwin Hubble used to discover the expansion of the universe.

Dr. Angel said part of the pleasure of the Giant Magellan project was working with old friends who could talk in shorthand.

In the mirror lab's air-conditioned conference room, Stephen Shectman from Carnegie said: "I've been coming here for 26 years. I thought I was done. But now we're just starting again."

Ever since Galileo's time astronomers have made telescope mirrors and lenses by grinding flat disks of glass together. But such rubbing produces a spherically shaped mirror that must then be re-shaped into a shallower curve known as a parabola — a delicate and error-prone process that has been the bane of many amateur and even professional astronomers. It was in the testing part of this part of the process, for example, that the builders of the Hubble Space Telescope stumbled, necessitating a dramatic series of spacewalks in 1993 to fit the orbiting telescope with corrective lenses.

Moreover, as mirrors have gotten bigger, the traditional method wastes a lot of glass and time. Dr. Angel estimated that about 20 tons of glass would have to be scooped out to make the new mirror. "That's a lot of glass at $40 a kilogram," he said.

Astronomers and physicists have long known, however, that the surface of a spinning liquid will form a parabola. Indeed, telescopes have been built using spinning pools of mercury as a mirror.

Dr. Angel, who was born in Lancashire, England, and earned a doctorate in physics at Oxford, said he was interested in astronomy as a child and once starting grinding a mirror for a telescope but never finished it. After completing his degree, in the late 1960's Dr. Angel went to work at Columbia, which was a center for the growing new field of X-ray astronomy, and he went along.

"I came into astronomy as an instrument builder," he said.

At the University of Arizona, where he moved in 1975, Dr. Angel started experimenting with casting mirrors in a kiln in his backyard in Tucson.

He found that by pouring the molten glass into a honeycomb structure he could make a mostly hollow mirror that was light, stiff and adjusted quickly to changes in the air temperature that would distort and disable fatter mirrors. Dr. Angel's friend and colleague Nick Woolf had emphasized these features as the key to building large telescopes in the future.

In 1985, a rotating furnace was installed under the football stadium, continuing a tradition in this country of important science projects under

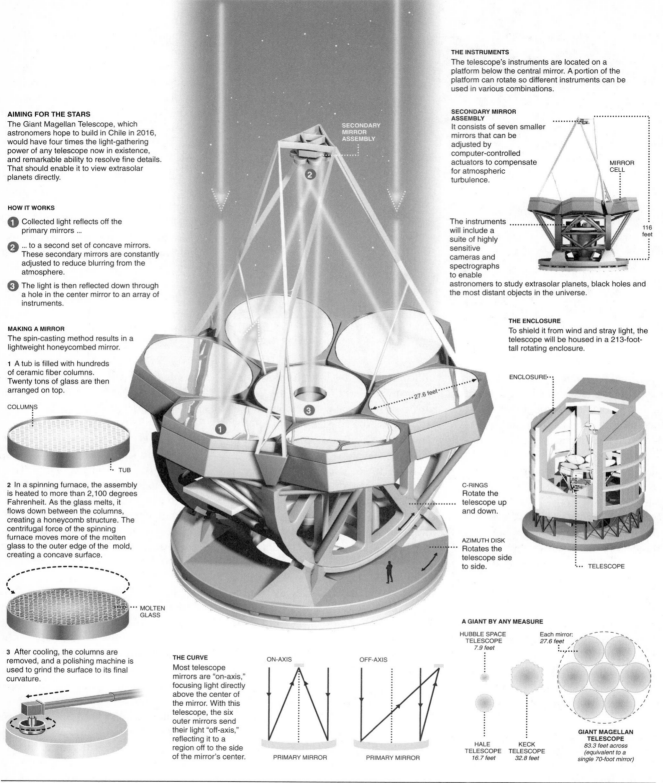

AIMING FOR THE STARS

The Giant Magellan Telescope, which astronomers hope to build in Chile in 2016, would have four times the light-gathering power of any telescope now in existence, and remarkable ability to resolve fine details. That should enable it to view extrasolar planets directly.

HOW IT WORKS

1 Collected light reflects off the primary mirrors ...

2 ... to a second set of concave mirrors. These secondary mirrors are constantly adjusted to reduce blurring from the atmosphere.

3 The light is then reflected down through a hole in the center mirror to an array of instruments.

MAKING A MIRROR

The spin-casting method results in a lightweight honeycombed mirror.

1 A tub is filled with hundreds of ceramic fiber columns. Twenty tons of glass are then arranged on top.

COLUMNS

TUB

2 In a spinning furnace, the assembly is heated to more than 2,100 degrees Fahrenheit. As the glass melts, it flows down between the columns, creating a honeycomb structure. The centrifugal force of the spinning furnace moves more of the molten glass to the outer edge of the mold, creating a concave surface.

MOLTEN GLASS

3 After cooling, the columns are removed, and a polishing machine is used to grind the surface to its final curvature.

THE CURVE

Most telescope mirrors are "on-axis," focusing light directly above the center of the mirror. With this telescope, the six outer mirrors send their light "off-axis," reflecting it to a region off to the side of the mirror's center.

ON-AXIS

OFF-AXIS

PRIMARY MIRROR

PRIMARY MIRROR

SECONDARY MIRROR ASSEMBLY

2

27.6 feet

3

1

C-RINGS
Rotate the telescope up and down.

AZIMUTH DISK
Rotates the telescope side to side.

THE INSTRUMENTS

The telescope's instruments are located on a platform below the central mirror. A portion of the platform can rotate so different instruments can be used in various combinations.

SECONDARY MIRROR ASSEMBLY

It consists of seven smaller mirrors that can be adjusted by computer-controlled actuators to compensate for atmospheric turbulence.

MIRROR CELL

116 feet

The instruments will include a suite of highly sensitive cameras and spectrographs to enable astronomers to study extrasolar planets, black holes and the most distant objects in the universe.

THE ENCLOSURE

To shield it from wind and stray light, the telescope will be housed in a 213-foot-tall rotating enclosure.

ENCLOSURE

TELESCOPE

A GIANT BY ANY MEASURE

HUBBLE SPACE TELESCOPE
7.9 feet

Each mirror:
27.6 feet

HALE TELESCOPE
16.7 feet

KECK TELESCOPE
32.8 feet

GIANT MAGELLAN TELESCOPE
83.3 feet across
(equivalent to a single 70-foot mirror)

Sources: Carnegie Observatories; Steward Observatory Mirror Lab; Paragon Engineering

Frank O'Connell/The New York Times

football stadiums. (The first controlled nuclear reaction occurred beneath the University of Chicago's old football stadium in 1942.) In 1990, it was enlarged to accommodate mirrors as large as 28 feet.

The present project grew out of the success of the Magellans, built by Carnegie, Harvard, Arizona, M.I.T. and Michigan, and completed in 2002. The Magellan astronomers like to brag about the quality of the images from these telescopes, which they attribute to the smoothness of the mirrors and the stability of the atmosphere at Las Campanas, where the Giant Magellan would be built.

"These mirrors work," Dr. Freedman said.

Astronomers say the Giant Magellan, augmented with so-called adaptive optics that reduce the blurring from the atmosphere, would be an invaluable tool, among other things, for hunting and studying planets around other stars.

"Bigger is better in a big way to see faint objects around bright ones," Dr. Angel said.

It could be years before the Giant Magellan group will know if their gamble has paid off. The mirror is scheduled to cool until late October, when technicians will pop the lid off the flying saucer oven and lift the new mirror out.

Only then will begin the arduous task of polishing and testing it. The testing is the big problem, Carnegie's Dr. Shectman said, "There's a schedule — two years, but . . .," he said, his voice trailing off.

And if after all this they don't get the money for the rest of the telescope?

"We didn't spend time on Plan B," Dr. Angel said.

As the Giant Magellan mirror was baking Dr. Angel and his colleagues led tours through what looked like a giant geek's playground. Towering into the shadows of one corner was a 400-ton tower for testing mirrors. Nearby, trailing a tangle of wires like a Medusa's head, hovered a thick aluminum disk the size of a small dinner table with padded feet.

That is the polisher, which can be bent under computer control as it strokes and shapes the mirror. To succeed, the mirror must be polished to within a millionth of an inch of the desired curve, Dr. Angel said.

A freshly polished 21-foot mirror built for Lockheed was sitting on its stand like a frozen whirlpool, honeycombs visible in its depths. Against the wall, hanging in a red steel frame, was a 28-foot mirror, the second of two that have been built for the Large Binocular Telescope on Mount Graham (the first is already in the telescope).

In still another corner was a polishing tool — a lump of curved granite a couple feet across covered with half-inch squares of black pitch — the same thing amateurs have been using forever.

"That's the great thing about this business," Dr. Angel said. "On the top it's high-tech, but at the bottom it's centuries old, two bodies rubbing against each other."

Dr. Angel's mirrors start off as white sand on the Florida Gulf Coast. It is cooked into borosilicate glass in one-ton batches in barrels at the Ohara glassworks in Japan and then smashed into smaller chunks.

Dr. Angel said the most important property of the borosilicate, besides low thermal expansion, was its uniformity, which allowed it to cool and harden without any built-in stresses that could later warp the mirror.

That and the fact that it melts at a finite temperature.

Dr. Angel's furnace generates two million watts. While they were casting their first 28-foot mirror, Dr. Angel explained, the furnace sprang a leak. The level of molten glass fell and kept falling. Some of it wound up on the floor, leaving part of the mirror's surface denuded.

The astronomers had to resort to what Dr. Angel calls a "crème brûlée process," adding more glass to the top of the mirror and firing the furnace up to remelt the top.

It worked, but the astronomers were still relieved on Giant Magellan casting day when the glass level held at the two-and-a-half-inch mark.

"It's good to check that off," Dr. Freedman said.

That night, the astronomers and their supporters gathered for a banquet and mused about the future.

Besides looking for alien planets, the Giant Magellan astronomers would like to investigate the so-called dark energy that seems to be splitting the universe apart.

But, as Dr. Kirshner noted, dark energy and extrasolar planets — the two big selling points for telescopes today — didn't exist 10 years ago and wouldn't have been on anybody's list of things to do. "You can't honestly predict what you're going to do," he said.

Patrick McCarthy of the Carnegie Observatories told the group, "The most important tool we take to the observatory is an open mind."

IN REVIEW

1. What is the Giant Magellan telescope? How many mirrors will it have, and how will they work together?

2. What is adaptive optics, and why will it be used with this and other future telescopes?

3. Describe a few of the scientific goals of the Giant Magellan telescope.

4. Briefly summarize some of the challenges of building large telescope mirrors, and how these challenges are being overcome.

With increasing technological capabilities, scientists have become inundated with data—so much, in fact, that it is increasingly difficult for them to analyze collected data, leaving the possibility that some important scientific truths may remain undiscovered. But scientists are turning back to technology to solve the problem of data overload. Using large databases and Internet software, scientists have created a virtual observatory, which acts as a large repository of information that is easily mined and analyzed to reveal new wonders of the universe. In addition to helping professional scientists, this virtual observatory has opened doors for amateur astronomers, in essence creating a new class of researchers that some call "citizen scientists." No longer an opportunity for only the elite, the virtual observatory and its software tools make science open to all who have a genuine interest.

Telescopes of the World, Unite!
A Cosmic Database Emerges

By Bruce Schechter
The New York Times, **May 20, 2003**

About a year ago a large group of astronomers began to assemble what some of them were calling "the world's best telescope." Their ambitious instrument is still far from complete, but they recently took it for a test run. Within minutes, to their joy and astonishment, they had discovered three or four brown dwarfs, objects that occupy the niche between planet and star.

"It gave me shivers when I heard about it," said Dr. Alex Szalay, a Johns Hopkins astronomer who is one of the telescope's chief architects.

It wasn't the brown dwarfs themselves that excited Dr. Szalay; hundreds of them have been discovered in the past decade. But he and many other astronomers believe that the means used to discover these objects heralds the beginning of a new era of astronomy, and even a new era of science.

The telescope that Dr. Szalay and his colleagues have constructed is not built of glass and metal. It is a virtual observatory, consisting of terabytes of data collected by dozens of telescopes on Earth and in space, and the software necessary to mine these data for scientific gems.

Like much of the rest of science, astronomy has been the beneficiary— and victim—of Moore's Law, which states that the capacity of computers and other silicon-based devices like charge-coupled devices, or C.C.D.'s, doubles every 18 months. (The C.C.D. has largely replaced photographic film in astronomical cameras.)

Projects like the National Virtual Observatory, which was created in response to the tsunami of data that is threatening to drown astronomers, is creating a new branch of science, Dr. Szalay believes.

Science, he points out, was "originally empirical, like Leonardo making wonderful drawings of nature." He continued: "Next came the theorists who tried to write down the equations that explained the observed behaviors, like Kepler or Einstein. Then, when we got to complex enough systems like the clustering of a million galaxies, there came computer simulations, the computational branch of science. Now we are getting into the data exploration part of science, which is kind of a little bit of them all."

Because its primary tools are computers rather than giant, multimillion telescopes, this new form of astronomy has the potential to democratize science.

"If at the same time most of the telescopes in the world are actually putting all of their data online with proper explanations," Dr. Szalay said, "then it doesn't matter where somebody is sitting, they can access all the data— either somebody in Baltimore, or somebody from Africa who got a Ph.D. in the U.S. and returned there and doesn't have access to a telescope but suddenly has a bunch of students. They can actually get to first-class data."

In the past 25 years the number of C.C.D. pixels in all the world's telescopes has increased by a factor of 3,000, with each pixel acting as a miniature astronomical instrument. The result, Dr. Szalay says, is that the total amount of astronomical data collected every year is doubling even while the amount spent on astronomy remains constant.

"We are getting overwhelmed," Dr. Szalay says. "With this explosion it's not just that individual telescopes are getting more and more data, but also the threshold gets lower, so that more and more groups are putting big cameras on their instruments. Even amateur astronomers today can generate gigabytes of data per night by attaching a digital camera to their telescope."

The problem is how to mine this vast store of data for the riches it almost certainly contains. Astronomers have been busy over the past couple of decades compiling complete surveys of the sky, encyclopedic catalogs of millions of astronomical objects viewed at many different wavelengths. These surveys exist in about 10 different spectral bands, from X-rays to the infrared, with each survey giving a different view of the universe.

The surveys contain about 100 terabytes (one terabyte is 1,000 gigabytes) of data, roughly five times as much as the Library of Congress holds. Unlike the Library of Congress, however, this information does not reside in a single place.

"There is no Library of Congress for astronomy," Dr. Szalay says, and as long as the data continue to accumulate at an exponential pace, there will never be one. Instead, "there will always be 8 or 10 large projects that contain 90 percent of the world's data at any one time."

The goal of the National Virtual Observatory is to make sure that "the current generation of professional and amateur astronomers are not overwhelmed by the chores of getting the actual data," Dr. Szalay says. "So we have to make it simple and easy for them to use the data in a friendly way."

In the first stage of the project this has meant creating tools that can search through different databases without requiring the searchers to be experts in their individual details. As a kind of shakedown cruise, the researchers at the National Virtual Observatory decided to focus on the data contained in two large sky surveys known as the Sloan Digital Sky Survey, which looks at the sky in the visible band of the spectrum and the Two Micron All Sky Survey, or 2MASS, which looks at the sky in the infrared.

"The reason we did those two is that they're very deep, they dig out objects that are very faint, much fainter than other surveys have been able to generate," said Dr. Bruce Berriman, a California Institute of Technology astronomer involved in the demonstration. "Because it goes to very faint objects you're able to dig out sources that are unusual or important in ways other projects can't do."

In particular, by combining the surveys they hoped to spot brown dwarfs. Brown dwarfs are essentially failed stars,

lumps of matter bigger than a planet but not large enough to kindle the thermonuclear fire of a star. As a result, they are relatively cool, emit very little light and are therefore difficult to spot.

The temperature of a star, like that of a glowing piece of metal, determines the color of light that it emits: the cooler the star the redder the light. The light from the brown dwarfs that the astronomers were searching for straddles the border between the infrared and the visible. This means that a brown dwarf should appear in the very shortest wavelength band of the infrared 2MASS survey and also in the longest wavelength band of the visible Sloan survey.

An astronomer looking at just the data from, say, the Sloan survey and seeing an object in a single band would probably dismiss it.

"Chances are pretty good that that single band detection is a piece of junk, some sort of artifact in the detectors in the telescope, a glint off a bright star, any number of things," said Dr. Davy Kirkpatrick, a member of the Caltech team. But if that same object also appears in the 2MASS data then the chances shoot up that it is something worth looking at more closely.

The astronomers developed a program that could access these different databases and search them for matches. Within a few minutes the computer spit out a half a dozen or so candidates for possible brown dwarfs. Most of these had been previously noticed in the data, which others had sifted through manually.

Finding these brown dwarfs was supposed to be the goal of the demonstration, a debugging run to prove that the software worked. But the computer also found several candidates for new brown dwarfs.

"Astronomers' first reaction when you find a new result is that there's something wrong," Dr. Berriman said. But after looking at the data more closely "it slowly dawned on us that this was something real, that this was a brown dwarf we found."

"Then our eyes started to widen up a little bit at the prospect of what might be coming in the future," he continued.

Dr. Szalay says, "This shows how many new things will come out in this process once hundreds of astronomers are using it all over the place."

What makes this result even more impressive is that the overlap of the two surveys covered something less than one two-hundredth of the sky, and yet they almost instantly found objects that astronomers, poring over data for weeks, had previously missed.

Another aspect of the National Virtual Observatory is the creation of an astronomical search engine, a kind of Google for astronomy that will allow amateurs and professionals to find astronomical resources.

"Were you to go to Google right now and type in the word galaxy, you wouldn't just get a whole bunch of astronomy sites," Dr. Berriman says. "You'd also find out about the L.A. Galaxy soccer team. There's even a town in Texas called Galaxy, with its own Web page. That's no good to astronomers because there's so much clutter in the sites."

To help solve this problem the National Virtual Observatory is creating an online registry of astronomical resources that should be available to the public early next year.

The registry, and indeed the entire virtual observatory project, is intended as a tool for anyone interested in astronomy.

"One of the major components of the registry is collecting information about the suitability of the resource for educational purposes, for amateurs, for students," said Dr. Robert Hanisch of the Space Telescope Science Institute in Baltimore. "We have a subgroup of our project concentrating on what kind of information does that clientele need to know."

The success of the National Virtual Observatory and similar projects means that exploring the heavens will no longer be limited to those few hearty individuals willing to sit freezing on mountaintops, waiting for the clouds to clear. Adding myriad seeking eyes and pondering brains to those already contemplating our place in the universe will be the greatest achievement of this new technology.

IN REVIEW

1. Explain why the amount of collected astronomical data has increased exponentially in the last few years.

2. How does the virtual observatory allow scientists and amateurs to analyze an increasing amount of collected data?

3. Explain why the virtual observatory could deepen understanding by elementary, middle, and high school students of the fundamental concepts of science.

4. How could data from the virtual observatory be used in science museums to generate interest by its visitors?

5. In your opinion, should astronomical data be analyzed by only those who have advanced degrees in science, or are there significant roles that amateur science enthusiasts can play? Do you know of any other tools besides the virtual observatory that allow amateurs to contribute to science?

Over the past couple of years, NASA has been developing plans to return humans to the Moon, with an eventual goal of building a lunar outpost and then sending humans on to Mars. Because we've been to the Moon before, we know that such plans are within our technological capabilities. The challenge is finding a way to do it within NASA's relatively tight budget constraints. This article describes the current plan for returning to the Moon.

NASA Planning Return to Moon Within 13 Years

By Warren E. Leary; Leslie Wayne contributed reporting from New York for this article.
The New York Times, September 20, 2005

WASHINGTON, Sept. 19—Combining an old concept, existing equipment and new ideas, NASA gave shape on Monday to President Bush's promise to send humans back to the Moon by the end of the next decade.

Michael D. Griffin, the agency's new administrator, detailed a $104 billion plan that he said would get astronauts to the Moon by 2018, serve as a steppingstone to Mars and beyond, and stay within NASA's existing budget.

The plan would use a new spacecraft similar to the Apollo command capsule of the original Moon program, and new rockets made up largely of components from the space shuttle program.

"It is very Apollo-like," Dr. Griffin said, "but bigger. Think of it as Apollo on steroids."

The plan drew a mixture of praise and criticism from lawmakers and space experts. [News analysis, Page A15.]

The chairman of the House Science Committee, Representative Sherwood Boehlert, Republican of New York, said it appeared to be "the safest, least expensive and most efficient way" of moving forward in space exploration, but added that current cost overruns in other NASA programs might make it hard to develop the new vehicle on schedule.

The outlines of the plan had been disclosed informally over the last two months by NASA officials and space experts. But Dr. Griffin's announcement laid out a timetable and a budget, putting flesh on the bones of a proposal that Mr. Bush announced in January 2004 but had never described in detail.

Photo by NASA, via Reuters

Another small step: A depiction of a proposed new lunar lander.

Dr. Griffin said that after adjusting for inflation, the program would cost just 55 percent of what it cost to put a dozen men on the lunar surface from 1969 to 1972.

The pay-as-you-go plan, approved by the White House last week, would stay within NASA's $16-billion-a-year budget through a combination of retiring the space shuttle, finishing the International Space Station and reallocating money from other NASA programs. And Dr. Griffin said the nation could well afford it, despite concern about tight budgets in the wake of Hurricane Katrina.

"We're talking about returning to the Moon in 2018," he said at a news conference here in Washington. "There will be a lot more hurricanes and a lot more other natural disasters to befall the United States and the world in that time, I hope none worse than Katrina.

"But the space program is a long-term investment in our future. We must deal with our short-term problems while not sacrificing our long-term investments in our future. When we have a hurricane, we don't cancel the Air Force. We don't cancel the Navy. And we're not going to cancel NASA."

Dr. Griffin and other space advocates, including influential members of Congress, have said the United States needs a replacement for the aging shuttle fleet as a matter of national security. Russia and China are currently capable of human spaceflight, and other countries have expressed interest in following suit. Dr. Griffin said the nation must maintain an independent capability to send people into space after the shuttle retires in 2010.

The new craft, called the crew exploration vehicle, would perch the astronauts' capsule above the rockets that power it into space, rather than alongside them as with the shuttle. NASA officials said it would be 10 times as safe as the shuttle, with a projected failure rate of 1 in 2,000, as opposed to 1 in 220 for the shuttle. The increased safety, they said, will be due in part to escape rockets that will be able to jettison the capsule away from the booster rocket in the event of an accident.

Back to the Moon

NASA has unveiled an ambitious plan to replace the shuttle and return Americans to the Moon by 2018. The design of the new craft is strikingly similar to the hardware used in the Apollo program nearly 50 years earlier, with a conical crew vehicle and a separate lander (not shown).

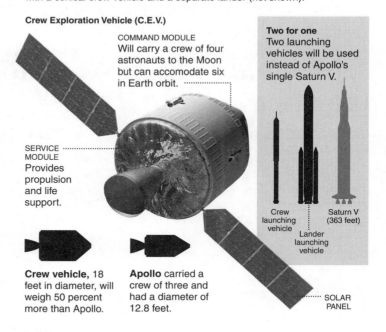

Crew Exploration Vehicle (C.E.V.)

COMMAND MODULE
Will carry a crew of four astronauts to the Moon but can accomodate six in Earth orbit.

SERVICE MODULE
Provides propulsion and life support.

Crew vehicle, 18 feet in diameter, will weigh 50 percent more than Apollo.

Apollo carried a crew of three and had a diameter of 12.8 feet.

Two for one
Two launching vehicles will be used instead of Apollo's single Saturn V.

Crew launching vehicle

Lander launching vehicle

Saturn V (363 feet)

SOLAR PANEL

From the Earth to the Moon
Unlike in Apollo, the new plan calls for the landing vehicle to be launched on a separate rocket from the crew vehicle. They will dock in Earth orbit before heading for the Moon. The entire crew will descend to the surface.

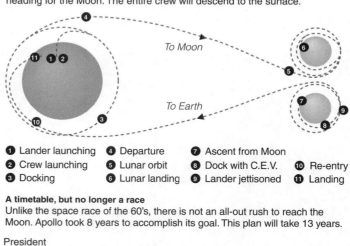

To Moon

To Earth

❶ Lander launching   ❹ Departure   ❼ Ascent from Moon
❷ Crew launching   ❺ Lunar orbit   ❽ Dock with C.E.V.   ❿ Re-entry
❸ Docking   ❻ Lunar landing   ❾ Lander jettisoned   ⓫ Landing

A timetable, but no longer a race
Unlike the space race of the 60's, there is not an all-out rush to reach the Moon. Apollo took 8 years to accomplish its goal. This plan will take 13 years.

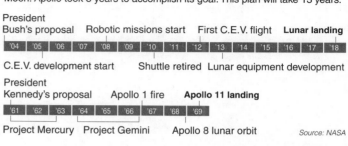

President
Bush's proposal   Robotic missions start   First C.E.V. flight   **Lunar landing**

| '04 | '05 | '06 | '07 | '08 | '09 | '10 | '11 | '12 | '13 | '14 | '15 | '16 | '17 | '18 |

C.E.V. development start   Shuttle retired Lunar equipment development

President
Kennedy's proposal   Apollo 1 fire   **Apollo 11 landing**

| '61 | '62 | '63 | '64 | '65 | '66 | '67 | '68 | '69 |

Project Mercury   Project Gemini   Apollo 8 lunar orbit

Source: NASA

The New York Times; C.E.V. image by John Frassanito and Associates

Dr. Griffin said the vehicle would be able to take as many as six astronauts to the space station, or fewer astronauts and some cargo. Or it could fly robotically without a crew, he added, carrying up some 25 tons of cargo, about as much as a shuttle can carry. The wingless craft, weighing 50 percent more than the Apollo, could carry a crew of four to the Moon.

As envisioned, the craft would resemble a larger version of the Russian spacecraft Soyuz, including solar-power panels deployed after launching. It would be carried aloft on a modified version of one of the shuttle's solid-fueled rocket boosters and a new second stage using one of the shuttle's main liquid-fueled engines. For Moon voyages, the craft would rendezvous in Earth orbit with lunar components lifted on a big new cargo rocket.

This heavy-cargo rocket, which could put 125 tons into orbit, would comprise two extended shuttle solid-fuel boosters attached to a liquid hydrogen-oxygen first stage made of an extended shuttle external fuel tank with five shuttle main engines. Atop this would be a new second stage using one or more of the shuttle main engines.

The bigger rocket, capable of lifting the payload of the Saturn V, which sent men to the Moon decades ago, would put into Earth orbit another rocket that could carry a landing craft and the crew vehicle to the Moon.

The Moon mission would be accomplished in stages. First, a vehicle with a lunar lander and a Moon rocket would be launched into Earth orbit. The crew capsule would be launched as much as a month later and would meet the first vehicle in orbit.

The joined craft would then be propelled to lunar orbit. From there, the landing craft would fly to the surface, and its crew of four would spend up to a week exploring.

Afterward, the top part of the two-piece lander would take the crew back to lunar orbit to rendezvous with the crew vehicle, which would have been left there unmanned, for the trip back to Earth. The bottom part of the lander could be left behind with equipment that might be used by future crews that land nearby, the agency said.

Nearing Earth, the crew vehicle would jettison its equipment module before the crew capsule plunged into the atmosphere. Parachutes would slow the capsule before it landed in the water or on land, possibly at Edwards Air Force Base in California. The landing would be cushioned by air bags on the craft's bottom or small rockets that fire just before touchdown.

Some of the big military contractors that stand to win multibillion-dollar contracts to launch and maintain the new mission vehicles said they were pleased by Monday's announcement.

"This is a big deal for us," said Michael Coats, chief of space exploration for the nation's largest military contractor, the Lockheed Martin Corporation, one of two teams competing to build the crew exploration vehicle. "Today's announcement means that we will have a stable program for a while. It would be great for us to compete and win a piece of the business."

Lockheed is the leader of a consortium that currently has a $28 million contract to design the crew exploration vehicle. Another team, headed by the Boeing Company and Northrop Grumman, has an identical contract to design an alternative. NASA is expected to select the winning design next spring.

John E. Pike, a space policy expert with the consulting firm GlobalSecurity.org, said NASA needed to sell more people on the value of a Moon program if it is to preserve future budgets and continue space exploration. The Moon may be a source of the rare element helium-3, for instance, which could fuel fusion reactors to provide abundant electricity on Earth, he said.

"NASA," Mr. Pike said, "needs to have reasons like this to go back to the Moon so the program has more stakeholders who see its value and want to protect it."

IN REVIEW

1. Briefly describe the vehicle being planned for the return to the Moon. How does it differ from the Apollo spacecraft that took humans to the Moon over three decades ago? How does it differ from the Space Shuttle?

2. Based on this article, summarize some of the arguments in support of returning humans to the Moon.

3. Do you think that NASA can succeed in this mission within the budgetary constraints discussed in the article? If so, why? If not, what do you think should be done? For example, should NASA's budget be increased?

4. While most scientists support the return to the Moon in principle, many fear that the expense of this new program will force NASA to cut important scientific missions. Already NASA has proposed cutting back on some of its science funding. Research the current status of NASA's science funding to find out how the new Moon program is impacting science.

5. Overall, do you support NASA's new plan? Defend your opinion.

Since December 1972 no person has set foot on the Moon, and despite successful follow-on missions of Skylab, the Space Shuttle, and the International Space Station, NASA's crowning achievement remains the Apollo program. The Apollo missions returned hundreds of pounds of rocks, greatly increasing our understanding of the geology of both the Moon and early Earth. However, this deepened geological knowledge of the Moon failed to capture the hearts and minds of the public. Now, with NASA proposing to return to the Moon, both the agency and the scientific community are struggling with selling this renewed exploration program to the American taxpayer. Specifically, what is the use of the Moon beyond geological investigation?

The Allure of an Outpost on the Moon

By Kenneth Chang
The New York Times, **January 13, 2004**

For some, it is the steppingstone of the Moon, not the distant goal of Mars, that is the irresistible destination in the human venture into space that President Bush will propose tomorrow.

For geologists, Moon rocks could tell much about the first billion years of Earth's history. For astronomers, the Moon would be a cold, dark place ideal for a telescope staring deep into the cosmos.

"There's a lot we could get out of the Moon," said Dr. Allan H. Treiman, a scientist at the Lunar and Planetary Institute in Houston. But he added, "It's not grab-the-public-by-the-throat science."

After the climactic triumph of Apollo, the public lost interest in continued human exploration of the Moon, and the Nixon administration cut deeply into NASA's budget. In the three decades since, NASA has focused robotic missions on more distant, more mysterious worlds like Mars and Jupiter and sent only two small orbiting spacecraft to the Moon. Clementine, a joint effort with the Department of Defense, found signs of frozen water at the lunar south pole in 1994. In 1998 and 1999, the space-on-a-budget Lunar Prospector mission, which cost $63 million, found even stronger evidence of ice and mapped out the Moon's gravitational and magnetic fields.

The primary reason to return and establish a permanent base on the Moon would be to assist a mission to Mars. Because the Moon's gravity is one-sixth of Earth's, gathering raw materials there—everything from metal for the spaceship to water for the astronauts to drink—would be much cheaper than hauling them up from Earth. So the cost and difficulty of traveling to Mars would be reduced.

The Moon base would also serve as a proving ground for new technologies developed for a Mars mission.

"If we learned to do that nearby, it might be a lot easier," Dr. Treiman said. "That's one of the big deals of using the Moon as a way station. We could learn how to live on Mars by living on the Moon."

But the Moon has its own appeal. About 4.45 billion years ago, a planetary interloper the size of Mars slammed into the infant Earth, tossing a blob of rock into space that became the Moon. With only one-eightieth Earth's mass, the Moon long ago cooled to the core, leaving it geologically dead. It is also too small to hold on to an atmosphere.

It is just this deadness that excites geologists. They see it as a museum of the history of the solar system. "To learn about the Moon is going to tell how the Earth formed," Dr. Treiman said.

On Earth, the continuous march of plate tectonics has destroyed almost all of the surface rocks from its first billion years. On the Moon, those rocks are still on the surface. The youngest rocks on the Moon are as old as some of the oldest rocks found on Earth: 3.2 billion years.

The craters on the Moon also preserve a record of the early bombardment of meteors. "The moon is a great place to study geological processes and it nicely complements the missing geological record on the Earth," said Dr. Paul D. Spudis, a planetary scientist at Johns Hopkins's Applied Physics Laboratory.

The 843 pounds of rocks brought back by Apollo astronauts revolutionized scientists' understanding of the Moon. The similar mix of oxygen atoms in the rocks of the Moon and Earth showed the two had a common ancestry instead of the Moon's forming elsewhere and then being captured by Earth's gravity. The chemical composition also showed there had never been significant amounts of water in most areas, except possibly the polar regions.

But the rocks came from just the six Apollo landing sites, leaving the rest of the surface, the size of Africa, unexplored. Dr. David S. McKay, a scientist at NASA's Johnson Space Center, would like the agency to go to other locations and dig trenches perhaps 100 yards deep to examine in detail the top layer of crushed rock and dust, known as the regolith.

"I think there is a lot of data hidden away on the Moon that remains to be unraveled," Dr. McKay said. "The lunar regolith is like a giant tape recorder that has been running for billions of years."

As astronomers try to look farther into the universe, they need a large telescope that can stay focused on a single patch of sky for weeks or months. Near absolute-zero temperatures and an airless environment are needed to prevent blurring. A nearby Moon base would allow easy repairs and upgrades.

"The Moon is a place that has been thought about a lot that has a lot of what you want," said Dr. J. Roger P. Angel, a professor of astronomy at the University of Arizona. He has proposed putting a large infrared telescope in a deep crater at the Moon's south pole.

At a Senate hearing on lunar exploration in November, Dr. Angel suggested that the mirror of such a telescope might consist of a round dish, 20 yards wide, with a reflecting liquid that spun at a rate of two revolutions a minute. The centrifugal force, coupled with the Moon's gravitational force, would push the liquid toward the outer edges of the dish to form a perfectly curved surface for gathering starlight.

Not only will a lunar telescope be more sensitive than the Hubble Space Telescope, but it should be able to detect galaxies and stars far fainter than will be seen by Hubble's planned replacement. It may even pick up light from the very first stars of the universe half a billion years after the Big Bang. "That's something you could do brilliantly from the Moon," Dr. Angel said.

IN REVIEW

1. In what major ways has past lunar exploration furthered our understanding of Earth?

2. List the major ways that a Moon base would assist in our exploration of Mars. Would a Moon base be better for future space exploration than an orbiting space station?

3. What would be the major difficulties in establishing a Moon base? How would our experience in living in Antarctic scientific stations help us? How would our experience in living in the International Space Station help us?

4. Explain why the Moon would be a good location for a large telescope. Why would astronomer Dr. Angel want to put a telescope that detects infrared light in a deep crater at the Moon's south pole? Explain.

5. In your opinion, would NASA's money be better spent on extending human exploration to the Moon and Mars, or should the agency focus its resources on robotic exploration?

Discoveries of extrasolar planets continue to pour in at an astonishing rate. And while early discoveries suggested solar systems much different from our own, some more recent ones suggest that solar systems like ours may prove to be common, after all.

Planet Group Similar to Solar System Is Found

By Dennis Overbye
The New York Times, May 18, 2006

A team of European astronomers said yesterday that they had found one of the closest analogues yet to our solar system: three planets and an asteroid belt circling a pale Sun-like star about 42 light-years away in the constellation Puppis.

The two innermost of the new Puppis planets, each about 10 times the mass of Earth, are probably rocky like our own home, but they circle too tightly about their star to be habitable.

The third planet, about 18 times the mass of Earth, circles at a distance of about 60 million miles, within the star's so-called habitable zone, where the temperatures would allow the existence of liquid water, the authors said.

The planet is still too big to be considered Earth-like, and is probably shrouded in hydrogen like Neptune or Uranus, and so is an unlikely environment for life. "Nevertheless," Christophe Lovis, of the University of Geneva and lead author of the group, wrote in an e-mail message, "this discovery opens the way to the detection of even smaller planets in the near future."

The discovery, to be published today in the journal Nature, continues a pattern of planet hunters leapfrogging one another to make discoveries that are increasingly alluring and suggestive of a life-friendly cosmos.

Charles Beichman, a planetary astronomer at the Jet Propulsion Laboratory in Pasadena, Calif., called the Puppis system "a wonderful planetary system," that people will be studying a lot over the next few years.

Like most of the 190 or so planets that have been discovered orbiting other stars, the Puppis planets were detected indirectly by the slight to-and-fro motion, or wobble, they induce in their home star, called HD 69830.

The first planets discovered in that way were giants, so-called hot Jupiters, hundreds of times the mass of Earth, circling very close to their stars, because proximity and large mass make for lots of wobble. As the measuring techniques have improved, the amount of wobble that can be discerned has shrunk, leading to the discovery of smaller Neptune-class planets, including one last year that was only seven times the mass of Earth.

Dr. Lovis and his colleagues at the European Southern Observatory at La Silla in Chile, used a 12-foot-diameter telescope equipped with a precise spectrograph to detect small wobbles in the Puppis star. It is the first system to be found in which all the planets are in the Neptune class.

Scientists knew there was something extraordinary about HD 69830, which is about 60 percent as luminous and 90 percent as massive as the Sun. Last year astronomers using the Spitzer Space Telescope reported that the star was surrounded by hot dust, probably the result of bumping and grinding in an asteroid belt 30 times more massive that the one in our solar system and the first one discovered around another Sun-like star.

Computer simulations by the European group suggest that the outermost planet orbits just outside the asteroid belt, where it would act as a gravitational shepherd to keep the belt in order.

Unfortunately for any prospective inhabitants of the Puppis system, however, the greater mass of the asteroid belt means planets there would be under constant bombardment, said Dr. Beichman, who was part of the Spitzer team. The Puppis system, he said, "would be a tough place to set up camp."

In a commentary accompanying the Nature paper, Prof. David Charbonneau of Harvard said it was a mystery how HD 69830 could have such a massive disk of debris around it while the star itself, according to the observations, has a lower abundance of heavy elements, the stuff of planets.

Dr. Charbonneau said the results were encouraging to those who hope to someday find Earth-like planets in habitable zones around other stars. In our solar system, that zone extends roughly from Venus to Mars and our planet induces too little wobble to be seen with present instruments. A smaller star would have a smaller habitable zone, and an Earth-like object there would have a bigger, more visible wobble effect.

If the European team's precision could be extended to such low-mass stars, Dr. Charbonneau wrote, "they might just turn up planets akin to our own."

IN REVIEW

1. Briefly describe the characteristics of the planetary system orbiting the star HD 69830. In what ways is this system more similar to our own than that of other known extrasolar planetary systems?

2. How were these planets discovered?

3. What are the best current guesses about the nature of the planets in this system? Do any of them seem likely to be habitable?

4. While we have not yet discovered Earth-like planets in any other planetary system, why does this new discovery make it seem more likely that such planets do indeed exist?

5. Comment on the likelihood that we will discover Earth-like planets in the near-future. How much effort do you think should be devoted to this search?

The planets of our own solar system are thought to have been born in a disk-shaped solar nebula, which formed as a natural part of the Sun's formation process. If so, such disks ought to be common around other stars as well, which is one reason why astronomers long suspected that planets would turn out to be plentiful in the galaxy and universe. Some of the first observational evidence for the formation of disks around other stars came in the 1980s, with observations of a system known as Beta Pictoris. This system continues to provide important new data to astronomers seeking to learn about the formation of stars and planets.

Second Disk Circling a Star May Provide Evidence of a Huge Hidden Planet

By John Noble Wilford
The New York Times, July 4, 2006

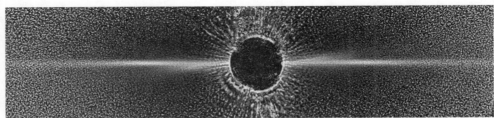

Photo by NASA/E.S.A.

Image of the star Beta Pictoris, taken by the Hubble.

Nearby and much younger and twice as massive as the Sun, Beta Pictoris is the bright star that cheered astronomers on to one of their biggest discoveries in recent time: finding planets around many other stars in the firmament.

Although planets have yet to be definitively detected around Beta Pictoris itself, the first observations of a luminous disk of dust there made scientists in the 1980's think they were seeing conditions like those existing in the planet-forming stage of the early solar system. So they were not too surprised a decade ago when signs of actual extra-solar planets elsewhere started popping up in their telescopes.

Now, Beta Pictoris has revealed another phenomenon that astronomers say may be a typical aspect of planet-forming processes. A fainter secondary dust disk circles the star, possibly betraying the presence of a planet up to 20 times as massive as Jupiter.

Astronomers reported last week that detailed photography by the Hubble Space Telescope showed the existence of two distinct disks around the star. The secondary disk is visible out to more than 24 billion miles from the star and is tilted by about four degrees from the main disk.

"This confirms what had been suspected before," said David A. Golimowski of Johns Hopkins University, a member of the discovery team. "The two disks suggest that planets could be forming there in two different planes, and this may even be the norm in the formative years of a star system."

The discovery was described in the June issue of The Astronomical Journal and in a telephone interview with Dr. Golimowski, the lead author of the journal report. The Space Telescope Science Institute in Baltimore said the new visible-light images were the best ever taken of Beta Pictoris, a relatively young star at less than 20 million years of age that is 63 light-years away in the southern constellation Pictor.

Previous Hubble photographs had revealed an odd warp in the Beta Pictoris disk, and astronomers speculated that this was actually a separate disk. Computer modeling by scientists at Grenoble Observatory in France offered a possible explanation. The gravitational force of a large planet in an inclined orbit could draw small planetary objects and rocks out of the main disk and align them in the planet's own plane.

Dr. Golimowski said this was still the best explanation, even though the planet organizing the second disk has so far eluded detection.

Astronomers suspect that multiple disks of orbiting dust may be common to certain young stars. Recent observations have picked up hints of a double disk around another star. Even the 4.6-billion-year-old Sun, when much younger, might have had planets forming on different planes, which Dr. Golimowski said could explain why the solar system planets are typically inclined to Earth's orbit by several degrees.

IN REVIEW

1. What is Beta Pictoris, and why was it famous even before the most recent discoveries?

2. Briefly summarize the latest discoveries about Beta Pictoris. What is their scientific significance?

3. How is computer modeling used in understanding observations such as those of Beta Pictoris?

4. What do these new discoveries tell us about the nebular theory of solar system formation?

The mission of the twin rovers Spirit and Opportunity now represents one of the greatest successes in the history of human space exploration. Designed to work only for three months, they are still going strong more than 2 1/2 years after their arrival on Mars.

Beyond Their Martian Dreams: Two Rovers Are Still Informing Experts Two Years Later

By Kenneth Chang
The New York Times, January 3, 2006

Two years ago today, the rover Spirit parachuted into the Martian atmosphere and, cocooned in protective air bags, bounced and rolled to a stop.

NASA's Jet Propulsion Laboratory erupted in cheers and hugs when the Spirit radioed word of its safe arrival. Three weeks later, there was more celebration as a second rover, the Opportunity, landed safely on the other side of Mars.

The plan was for each rover to explore for 90 Martian days (each Martian day, called a Sol, is almost 40 minutes longer than an Earth day). But once they rolled off their platforms and began work, mission managers knew that the vehicles might survive longer than 90 days.

"I really thought we would get 120," said Steven W. Squyres, a professor of astronomy at Cornell and the mission's principal investigator. "Maybe 150, and absolute maximum 180."

In other words, Dr. Squyres thought that by a year and a half ago, he would be devoting all his time to analyzing the trove of data collected. Instead, the Spirit's jaunt has continued past its 700th Martian day. The Opportunity is also doing well. And Dr. Squyres still spends much of his time figuring out where to send them.

Each has suffered some decay. The Spirit has worn out the diamond bits on its drill, and the Opportunity has a balky shoulder joint on the arm that holds most of its instruments, and one of its turning motors has failed. But neither seems close to demise.

The successes rose out of the shambles of NASA's Mars program in 1999. A failure to convert between metric and English units caused the Mars Climate Orbiter to burn up in the atmosphere. The Mars Polar Lander disappeared as it approached the surface; a design flaw is believed to have caused the lander to shut off its engine too early, causing it to crash.

NASA scrapped everything on its Mars planning board except for the Mars Odyssey orbiter, already built and nearly ready for launching, and started almost from scratch with the two rovers, building on successful technology that the agency had used for its 1997 Pathfinder mission.

The schedule in 2003 was tight, but unlike the 1999 failures — pushed too far by the "better, faster cheaper" philosophy of Daniel S. Goldin, then NASA's administrator — the rover project received robust resources.

"It had to work, and everybody was behind it," said Barry Goldstein, who was the deputy flight system manager on the rover project before moving on to another Mars mission. "We got the best people. We worked very hard to understand what could happen and how to protect against them."

For example, tests revealed that the parachutes fluttered uselessly instead of opening, and the fabric of the air bags, which had worked flawlessly in Pathfinder, shredded because of the heavier weight of the rovers. The air bags were strengthened with fabric, and the parachute design was tweaked.

The rovers have also benefited from having been designed for worst cases. Many of the actual conditions have turned out to be not nearly as severe. Temperature swings between day and night on Mars contract and expand the rovers' motors and electronics, for example, and eventually parts become loose and pop off.

Not knowing exactly what the conditions would be like, the engineers designed systems that could take a daily swing of 160 degrees. Actual swings experienced by the rovers' internal systems have usually been 50 to 70 degrees.

The rovers also had some luck. The accumulation of dust on the solar panels was expected to cut the power output to levels below what would be needed to survive the winter. The dust did accumulate but was cleaned off when micro-tornadoes known as dust devils swept over both rovers. Suddenly the panels returned nearly to their original power output.

As the rovers kept going and going, a group of engineers, once worried about the worst case, pondered the best case.

They reviewed the test results for the motors, electronics and other systems on the rovers to estimate when they would wear out. "At that point, we came to the conclusion there's a reasonable chance that these things could live two to four years," said Daniel Limonadi, a flight system engineer who was on that committee. Current financing continues through next September.

Above, the Spirit's view of Gusev Crater in September; below, the Opportunity's recent images of the Erebus Crater.

"On the one hand, we're tired," Dr. Squyres said. "On the other hand, there is no thrill in science that matches the thrill of daily discovery. And we've been doing that for two years now."

Two Earth years equal just more than one Mars year, which means the rovers have now seen one full cycle of seasons. "Once you've lived with these rovers for a full Martian year, you kind of get a feel for what the planet's like," Dr. Squyres said. "You just develop a deeper understanding of what it would be like to be on Mars."

At the Opportunity's site, the skies turned cloudier in winter and became clear again with spring. Meanwhile, dust devils swirled across the landscape most often during summer, especially at the Spirit's site.

Of late, the exploring has been particularly fruitful for the Spirit, which has been in a series of hills within 95-mile-wide Gusev Crater. In the plains where the Spirit landed, every rock monotonously proved to be basaltic lava. But in the hills, which are believed to consist of older rocks, the Spirit has found more than 10 types of rocks, and many show signs of alteration by small amounts of water.

What the Spirit has not found, though, are signs of a lake that is thought to have filled the crater long ago. (Photographs from orbit show a channel leading to the crater.)

"We have seen absolutely no evidence for surface water," Dr. Squyres said. He says he still thinks that the plains of Gusev were under water at some point and that the hills were islands then. But volcanoes probably buried the lake bottom under lava, along with the rocks that would show evidence of the lake, he said.

If two years of roving have shown anything, it is that the exploration will never be complete. "Whenever the rovers die," Dr. Squyres said, "tomorrow or two years from now, there will always be something wonderful and tantalizing just beyond our reach that we will never get to."

IN REVIEW

1. How did the failures of past Mars missions contribute to the success of the rovers?

2. What factors have played a role in allowing the rovers to last so long?

3. How have the extended lives of the rovers enabled scientists to deepen human understanding of Mars?

4. What advantages do robotic explorers, like the rovers, have over human explorers? What advantages do human explorers have over robots?

5. The total program cost (including construction, launch, and operations) for the two Mars rovers has been close to $1 billion. How much is this per person in the United States? Do you think the cost is justified by the science return? Defend your opinion.

Mars is a frigid planet with an extremely thin atmosphere and no liquid water on its surface. Yet, of all the planets in the solar system, Mars is the planet most like our home. Its dry land area is roughly equal to that of Earth, and like Earth, Mars has volcanoes, canyons, dry riverbeds, and polar caps of water ice. Indeed, one of the most important tools that scientists use to understand Mars is direct comparisons with Earth. The Mars exploration rovers, which have returned several daily images from the surface of the planet, have provided scientists with a huge amount of data. By comparing these images to what we know from the Earth's surface, scientists have been able to deepen their understanding of Martian geological history.

Mars' Round, Smooth Stones Have a Counterpart in Utah

By Kenneth Chang
The New York Times, June 22, 2004

Among the most intriguing discoveries from Mars this year are the almost perfectly round pebbles, nicknamed blueberries, that the rover Opportunity found strewn about at its landing site.

Similar spherical pebbles, it turns out, are also strewn around southern Utah.

To better understand the geological history of Meridiani Planum, the vast plain near the Martian equator that Opportunity is currently exploring, scientists are looking for places on Earth that might have formed under comparable conditions.

Researchers from the University of Utah, the National Institute of Aerospace Technology in Madrid, and G. d'Annunzio University in Italy found the Utah pebbles in several places, including Zion and Capitol Reef National Parks. They say the Utah pebbles, which formed about 25 million years ago when iron-rich groundwater percolated through porous sandstone, provide more evidence that the Meridiani Planum region was once drenched in water and perhaps amenable to life.

"It's hard to generate that kind of shape with other mechanisms," said Dr. Marjorie A. Chan, chairwoman of the geology and geophysics department at the University of Utah and the lead author of an article in the current issue of Nature that compares the Utah pebbles with their Martian counterparts.

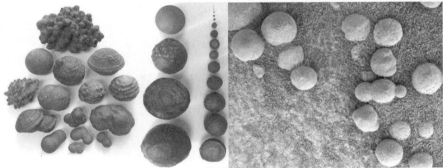

Photos by: left, Marjorie Chan, Brenda Beitler/University of Utah; right, NASA/JPL

Spherical pebbles strewn around southern Utah, left, which formed about 25 million years ago, are similar to the round pebbles found this year on Mars, right.

Planetary scientists chose Meridiani Planum as the destination for the Opportunity rover because it was there that orbiting spacecraft had spotted a vast deposit of hematite, an iron mineral that usually forms in the presence of water. From orbit, however, scientists could not tell which rocks contained the hematite or how the hematite got there.

Dr. Chan said that before Opportunity landed, she had predicted that the hematite would be found in spheres like those she had studied in Utah for eight years. When Opportunity did indeed find blueberries, "we were just amazed, oh my gosh it's true," she said.

The Utah blueberries were scattered on the surface and also embedded within bedrock. On Mars, the rover's probes showed that the blueberries, but not the surrounding rock or soil, contained bountiful amounts of hematite.

Scientists working on the rover mission concluded that the blueberries were objects known as concretions. Concretions form out of minerals that precipitate out of iron-rich water into surrounding sedimentary rock; because the composition of the rocks was uniform, the concretions grew equally in all directions into spheres.

Dr. Scott M. McLennan, a professor of geosciences at the State University of New York at Stony Brook and a member of the mission science team, said the Utah findings were interesting and potentially useful. "When you look at the blueberries concentrated on the surface on Utah," he said, "they look remarkably similar to the concentrations of blueberries on the Martian surface."

The Utah pebbles are not a perfect match, however. They come in a wider range of sizes, as large as grapefruits or even basketballs; the ones on Mars range from the size of BB's to small blueberries, all less than a quarter inch in diameter. That may be because Opportunity has examined only a small area, or it may reflect a difference in how the concretions formed.

In other respects, Meridiani Planum more resembles other places on Earth. On Mars, the surrounding sedimentary rock is made of ground-up volcanic basalt like that found on the beaches of Hawaii; the Utah concretions formed within sandstone. The water at Meridiani Planum was also highly acidic, like the Rio Tinto in southwestern Spain, whose name translates as Red River from the rusty hue it acquired from iron washing out of the rocks.

Dr. McLennan that geologists would have to look at many places on Earth to piece together the story of what happened at Meridiani Planum billions of years ago. "My guess," he said, "is that's going to be hard to find a perfect analog."

IN REVIEW

1. List how scientists have used comparisons with Earth to better understand the geological history of Meridiani Planum.

2. How do the spherical pebbles imaged on Mars provide evidence of past persistent liquid water on the surface of the planet?

3. Explain how spacecraft that orbit and/or land on Mars complement each other to aid scientists in understanding the planet, particularly with regard to the issue of minerals that form in water.

4. What are the significant differences between the spherical pebbles imaged on Meridiani Planum and those found in Utah? In your opinion, do these differences invalidate the scientists' conclusions about past persistent surface water on Mars?

5. How might human exploration of Mars provide more efficient and concrete data than those provided by the robotic rovers?

There is no liquid water on the surface of Mars today. As discussed in your text, we know this not only because we haven't seen any, but also because the atmospheric pressure is too low for liquid water to remain stable. Nevertheless, abundant evidence points to the idea that Mars had surface liquid water in the distant past. This article discusses geological evidence for past surface water collected by the Opportunity rover.

Scientists Report Evidence of Saltwater Pools on Mars

By Warren E. Leary
The New York Times, March 24, 2004

Mars was once a much warmer, wetter place, with pools of saltwater that sometimes flowed across the surface, scientists reported Tuesday.

Analyzing findings from sedimentary rocks explored by the rover Opportunity, the scientists said the rocks now appeared to have formed under a shallow bed of softly flowing water near a shoreline — not, as formerly seemed possible, through seepage from underground.

It was the first concrete evidence that water might have flowed on the Martian surface, and it provided new hints that life may have existed there.

"We think Opportunity is now parked on what was once the shoreline of a salty sea on Mars," Dr. Steven W. Squyres of Cornell University, principal investigator for the science payload on the Opportunity and its twin Mars exploration rover, Spirit, said at a news conference here at NASA headquarters.

"If we are correct in our interpretation, this was a habitable environment," Dr. Squyres went on. "It's a salt flat. These are the kinds of environments that are very suitable for life."

He said there was still no evidence that life existed at the site, and the Opportunity does not have the instruments to hunt for microscopic fossils. But rock formed from sediments from evaporating standing water, like those at the site, "offers excellent capability for preserving evidence of any biochemical or biological material that may have been in the water," he said.

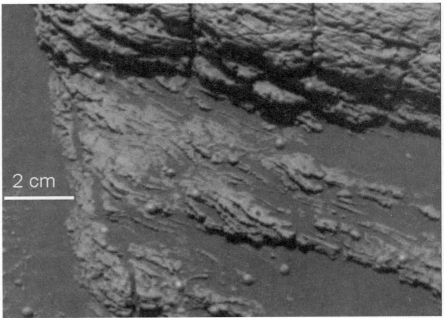

2 cm

Photo by NASA, via Associated Press

This image of a rock taken by a Mars rover is said to provide evidence that sediments forming the rock were laid down by flowing water.

The rover Opportunity landed on a plain called Meridiani Planum on Jan. 25, three weeks after the Spirit touched down on the opposite side of the planet at a site called Gusev Crater. NASA officials said Tuesday that the two craft were operating so well that their original 90-day missions would very likely extend into the summer.

Dr. Edward Weiler, NASA's associate administrator for space science, said that when researchers first announced on March 2 that the Opportunity had found rocks that formed in the presence of water, there was some indication that the water pooled on the surface. But to be sure that they were interpreting the data correctly, Dr. Weiler said, they sent it to several outside experts for review before Tuesday's announcement.

One of those scientists, Dr. David Rubin, a sedimentary expert with the United States Geological Survey in Santa Cruz, Calif., said he was shocked when he received the Mars pictures from NASA. "I was astonished," Dr. Rubin said. "They looked like sedimentary deposits found at a beach on Earth."

Dr. Rubin told the news conference that some of the pictures showed rocks that could have been formed by particles deposited by the wind, and others by sediments put down by water. However, he said, there are features in some of the formations that require flowing water, which supports the explanations by the rover scientists.

Dr. John Grotzinger of the Massachusetts Institute of Technology, a member of the rover science team, said the rocky outcroppings examined by the Opportunity have ripple patterns and salt concentrations that are telltale signs of rock formed in standing water that sometimes flowed. This would be like some salt flats on Earth, which are periodically submerged, Dr. Grotzinger said.

To examine the patterns in the Mars outcroppings, the Opportunity used a microscopic camera on its robot arm to take multiple postage-stamp-sized pictures that were put together into mosaics to examine the layering in the rock.

The close-up pictures revealed that the sediments that bonded to form the rocks were in uneven layers distorted by the ripples of flowing water, patterns called crossbedding and festooning, Dr. Grotzinger said.

Patterns in some of the layered rocks indicate they were shaped by ripples at least two inches deep, and possibly much deeper, flowing at a speed of 4 to 20 inches per second, he said.

The patterns include distinctive, upward-turning curves characteristic of water depositing particles in layers rather than wind, he said.

In further evidence of water formation, Dr. Grotzinger said, the rover found chlorine and bromine salts in the rocks, suggesting that salt concentrations were rising while water was evaporating.

Dr. Weiler said the findings would spur efforts to expand Mars exploration and look for evidence of past or current life. He noted that NASA planned to send the Mars Science Laboratory, a more sophisticated rover that will do detailed life-science testing, to the planet in 2009 and that Meridiani was now a prime site.

To search for further evidence of Martian water, the Opportunity has climbed out of its landing site and will head for a crater named Endurance, 2,300 feet away. That crater, scientists said, may have even deeper rock formations.

IN REVIEW

1. What are the major lines of evidence that liquid saltwater once existed at Opportunity's landing site?

2. Explain how NASA assured that their observations and theories about past liquid water on Mars were robust and sound.

3. What scientific measurements did the Opportunity rover make to enable scientists to make their conclusions about the landing site?

4. Might the rover's findings affect future missions to Mars? In your opinion, should NASA continue to explore Mars given the fact that liquid oceans may have dried up over 3 billion years ago? Explain.

The atmosphere of Mars is very thin, just 1% of the density of Earth's. Made up mostly of carbon dioxide, the Martian air contains virtually no oxygen. With this lack of oxygen and very low pressures, Mars' atmosphere is toxic to humans, and future astronauts will require special suits as they roam the surface. Despite these differences, there are some key similarities between the atmospheres of Earth and Mars. For example, Earth and Mars have clouds and fog consisting of water ice and vapor. Recently, scientists have discovered another likeness of Mars to Earth: trace amounts of methane in the air of Mars, a potential indicator of life below the surface.

Methane in Martian Air Suggests Life Beneath the Surface

By Kenneth Chang
The New York Times, November 23, 2004

A third team of scientists has now reported a seemingly simple discovery on Mars: its atmosphere contains methane.

But that finding has potentially profound implications, including the possibility of present-day microbes living on Mars.

Speaking this month at the American Astronomical Society's Division for Planetary Sciences meeting in Louisville, Ky., Dr. Michael Mumma, a senior scientist at NASA's Goddard Space Flight Center in Greenbelt, Md., reported three years of observations had provided strong evidence for methane.

"We are 99 percent confident," Dr. Mumma said. "It surprised all of us, actually. We really are still scrambling to understand what it means."

Methane, the simplest of hydrocarbon molecules with one carbon and four hydrogen atoms, is fragile in air and easily broken apart when hit by ultraviolet light. Calculations indicate that any methane in the Martian air must have been put there within the past 300 years.

That then raises the question: What is putting methane into the Martian air?

There seem to be only two plausible explanations. One is geothermal chemical reactions involving water and heat like those that occur on Earth in the hot springs of Yellowstone or at hydrothermal vents on the bottoms of oceans.

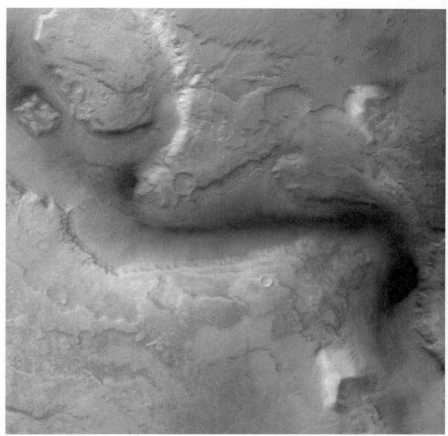

Photo by European Space Agency via Associated Press

Meandering river channels like this in Mars' southern hemisphere helped give rise to the notion that the planet had once been a tropical paradise, warmed by a heat-trapping blanket of carbon dioxide in its atmosphere.

That would intrigue planetary geologists. Although frozen water is known to exist, there are no signs that any volcanism has occurred there for millions of years. Also, an instrument aboard NASA's Mars Odyssey looked for warm spots on Mars' surface and did not find any.

The other, more intriguing, is life. On Earth, a class of bacteria known as methanogens breathes out methane as a waste product. The discovery, if confirmed, suggests that perhaps Martian life arose on a presumably more hospitable Mars billions of years ago and survives to this day underground, beneath the cold, dry landscape.

Dr. Vladimir Krasnopolsky of Catholic University in Washington, the leader of one of the teams, said he believed bacteria to be the "most plausible source."

Others are more cautious. "Three difficult detections, or marginal detections, don't equate to one really strong one," said Dr. Philip R. Christensen, a professor of geological sciences at Arizona State University.

Dr. Krasnopolsky's findings, relying on observations from the Canada-France-Hawaii Telescope in Hawaii, were first reported at a conference in Europe this year and will be published in the journal Icarus.

In January, scientists working on the European Space Agency's Mars Express mission also reported the detection of the methane. A few months later, that group, led by Dr. Vittorio Formisano of the Institute of Physics and Interplanetary Science in Rome reported that the methane appeared to be more plentiful in regions where frozen water is known to exist underground.

All three teams of astronomers looked for methane molecules in the Martian air by examining the rainbow of light reflected by the planet. Different molecules absorb different, very specific colors, producing a bar-code-like series of black lines blotting out part of the rainbow spectrum. The widths of the lines tell the quantity. Dr. Krasnopolsky and Dr. Formisano based their claims on a single dark line.

The journal Science published the Mars Express results this month. Dr. Christensen of Arizona State said he was unconvinced by it. "I must confess I'm surprised it was published," he said. "I think it's just instrument noise. This detection is right at the noise level of the instrument."

Dr. Mumma said his ground-based observations from Hawaii and Chile spotted two separate dark lines corresponding to methane and performed other checks. "Mike's a really careful guy," said Dr. Steven W. Squyres, principal investigator for the rovers now on Mars, who attended Dr. Mumma's talk. "It was to me, by a significant margin, the most compelling argument that I've seen."

There is a new wrinkle in Dr. Mumma's findings: some regions of Mars near the equator possess surprisingly high levels of methane, up to 250 parts per billion, while areas near the poles had 20 to 60 parts per billion. Earth air, by comparison, contains about 1,700 parts per billion of methane. Dr. Mumma's readings are considerably higher than those reported by the other two groups.

Scientists have generally thought that methane, if present, would quickly distribute evenly through the atmosphere, so the clumps of high concentration suggest that not only are there sources emitting methane, but perhaps some process is also destroying methane over the poles.

The methane findings on current-day Mars come as planetary scientists are again rethinking their ideas about long-ago Mars. Geological carvings on the surface, from ones that look like meandering river channels to gigantic canyons, gave rise to the notion that Mars had been a tropical paradise, perhaps warmed by a thick heat-trapping blanket of carbon dioxide in its atmosphere.

But climatologists found that it was hard for their computer models to provide that much warming, and scientists shifted to a picture of Mars as wet, but cold. Many of the features could have been cut by glaciers or transitory hellish deluges when ice was melted by meteor strikes.

Mars also possesses few carbonates, the minerals in limestone that would be expected to form in the presence of water, but does have much olivine, a mineral that falls apart when exposed to moisture.

This year, however, the rover Opportunity, which landed at a site called Meridiani Planum, found minerals and salts that indicate that that part of Mars at least had once been soaked in water, although when and for how long remain uncertain. Dr. Squyres also noted that while the minerals indicate liquid water, "We see nothing that looks like wave ripples" in the layers of sediments preserved in the rocks.

The other rover, Spirit, on the other side of Mars, initially found only volcanic rocks that appear almost unchanged for billions of years. It has since rolled to nearby hills, which appear to be slightly older, where the rocks seem to have been significantly changed by water.

The rover findings and others presented last month in Jackson Hole, Wyo., at a conference about early Mars have led some to think again of the planet long ago as warm and wet.

Even Dr. James F. Kasting, a climatologist at Penn State whose models helped convince people that Mars had not been warm, has changed his mind. Dr. Kasting is now investigating methane, a more potent greenhouse gas than carbon dioxide, as a cause of warming. His initial simulations show methane cooling the planet but he thinks the error is in his calculations, not his hypothesis.

"I think it's our problem, not Mars' problem," he said. "I think the evidence keeps mounting that it was warm. I think it has to be stably warm."

The opinion is not unanimous, but the idea of early oceans is gaining favor. Some scientists, like Dr. Stephen M. Clifford of the Lunar and Planetary Institute in Houston, said that four billion years ago the decay of radioactive elements in the core of Mars would have produced enough heat to melt ice from below, producing an ice-covered ocean. Acidic waters could explain the lack of carbonates.

Dr. Daniel J. McCleese, chief scientist for Mars exploration at the NASA's Jet Propulsion Laboratory, said that during discussions someone said, "So we all believe there were oceans on early Mars?"

Dr. McCleese said: "Nobody spoke against that. Then someone said, 'What about a warm climate?' And then a tumultuous exchange began."

IN REVIEW

1. Explain why methane would be fragile in the Martian air.

2. According to scientists, what are the two possible sources of methane? What are the major reasons that scientists have proposed that these are the possible sources? List the major reasons some scientists are cautious about these hypotheses.

3. Explain how data from orbiting spacecraft and Earth-based telescopes are providing a stronger line of evidence for methane in the Martian atmosphere.

4. What are the implications of finding spotty regions with higher concentrations of atmospheric methane?

5. How might methane help explain the past climate of Mars? Why is there still uncertainty about how warm Mars was in the past?

Your textbook describes the spectacular success of the Huygens probe in reaching the surface of Saturn's moon Titan, and early analysis of the probe's findings. The exploration of Titan still continues, both with further analysis of the Huygens data and through many more close passes of Titan by the Cassini spacecraft that currently orbits Saturn. This article describes some of the recent findings, including the fact that methane is apparently being replenished in the atmosphere by an internal geological source.

Probes Reveal Methane Haze On a Dynamic Saturn Moon

By Warren E. Leary
The New York Times, December 1, 2005

WASHINGTON, Nov. 30 — Robotic explorers probing neighboring planets have found evidence of hidden impact craters on Mars and dynamic weather, possibly including lightning, on Saturn's giant moon Titan, scientists reported on Wednesday.

In a series of papers being published this week in two scientific journals, the scientists report that smog-shrouded Titan is a frigid, dynamic world of ice carved and colored by liquid methane and organic chemicals.

The European Space Agency's Huygens mission, which landed on Titan on Jan. 14 after a seven-year ride to Saturn on NASA's Cassini spacecraft, parachuted through winds of up to 280 miles an hour while 75 miles above the surface. At a news conference in Paris, the researchers said the winds decreased at lower altitudes and dropped to walking speed at the surface.

On its descent of 2 hours 28 minutes to the minus-290-degree surface, the Huygens craft found a surprising electrically charged ionospheric layer bearing evidence of lightning, roughly from 85 miles to 25 miles above Titan's surface, the researchers reported in papers published online and in the journal Nature.

The layer of haze of methane and suspended particles that surrounds the planet-size moon goes all the way down to the surface, the spacecraft found, contrary to many predictions. But the haze cleared enough below 25 miles for the craft to take clear pictures of an ice world eroded by liquid flows and studded with wide flat areas that could be the remnants of lakebeds once filled with liquid methane.

"The surface is very dark," Bruno Bezard of the Paris Observatory said. "It would appear brownish to the human eye. What we see is a kind of dirty water ice mixed with other components."

Jonathan Lunine of the University of Arizona said radar images made by the Cassini craft when it passed by Titan suggested a strange world of rough lighter-colored areas and smooth dark stretches. The Cassini radar, which can penetrate the haze, has mapped 2 percent of the hidden surface of Titan, and experts said they hoped it would detail 10 percent to 20 percent on 37 more passes planned over the next four years.

A NASA-built instrument on the Huygens craft analyzed atmospheric gases in its descent and found that the source of the methane appeared to be the interior of Titan. Biological processes or degrading organic material can produce methane, the primary component of natural gas on Earth.

"We have determined that Titan's methane is not of biological origin, so it must be replenished by geologic processes on Titan," said Hasso Niemann of the Goddard Space Flight Center, a principal NASA investigator for the instrument.

IN REVIEW

1. What would the surface of Titan look like to the human eye?

2. Did scientists expect to find methane haze on Titan? Explain.

3. What fraction of Titan's surface had been mapped by Cassini radar as of late 2005?

4. How will we acquire additional data about Titan over the next few years?

Your textbook section on Titan concludes with a paragraph that begins, "Perhaps the most astonishing result from the Huygens mission is how familiar the landscape looks in this alien environment. . ." The dunes described in this article are yet another discovery that emphasizes how Titan's very different surface materials nonetheless produce features that look quite similar to surface features on Earth.

NASA Images Give New View of a Saturn Moon: No Oceans, but a Sea of Sand

By Kenneth Chang
The New York Times, May 9, 2006

A few years ago, many planetary scientists entertained visions of Titan, a Saturn moon, awash in oceans — not of water, but of ethane, a hydrocarbon gas that can condense to liquid at the surface, where the temperatures average minus-290 degrees Fahrenheit.

But images from NASA's Cassini spacecraft now show the opposite picture: the equatorial regions appear arid, with vast seas of sand dunes like those in the Sahara. That is all the more remarkable because Titan almost certainly does not have sand like the sand in the Sahara.

The radar images, reported in the current issue of the journal Science, show parallel, east-west lines of dunes, which are spaced about a mile apart, reach more than 300 feet high and run up to 930 miles long.

The dunes raise two questions: What are the grains made of? What is creating the wind to blow them around?

Because Titan is so far from the Sun, scientists had expected the air there to be still, with little energy from sunlight to power the winds. Now, however, they think gravitational tides produced by Titan's elliptical orbit around Saturn may be enough to drive winds averaging one mile per hour, and gusts would be faster. With Titan's lesser gravity and thicker air, that is enough to blow the sand around.

As for the sand itself, Titan's grains are most likely bits of water ice or organic solids, not silicates like the sand on Earth. (Titan is roughly half rock and half ice, and the rocky material like silicates probably sank to the moon's middle.)

Despite the very different conditions on Titan, the parallel lines of dunes look virtually the same as those in the Sahara or in the Namib Desert in Africa.

"Somehow all this stuff cancels out," said Ralph D. Lorenz, a scientist at the University of Arizona and an author of the Science paper, "and it gives you the same landscape you find on Earth."

Photo by Science

Photo by Earth Sciences and Image Analysis Laboratory/NASA

An image of Titan, a moon of Saturn, left, shows lines of dunes like those in the Namib Desert in Africa, right.

IN REVIEW

1. Describe the basic appearance of the dunes observed on Titan.

2. Why didn't scientists expect wind on Titan, and how do they now explain the existing winds? How do wind speeds on Titan compare to those on Earth?

3. What is the likely composition of the grains that make up the dunes on Titan?

4. Overall, what lessons does Titan teach us about the types of geological surface features that can be produced on different worlds? Explain your opinion clearly.

In an interesting cycle, scientists, engineers, and science-fiction writers influence each other in ways that both lead to a deeper understanding of the universe and expand our imagination beyond our current knowledge level. For example, the theories of Albert Einstein and Stephen Hawking have been used by science-fiction writers to explain how characters travel across extreme distances of space and time. Conversely, groups of scientists and engineers are researching the possibility of greater-than-light-speed travel using ideas proposed by the *Star Trek* and *Star Wars* series. One such scientist who has been strongly linked to science fiction is Lawrence Krauss, an internationally known theoretical physicist who wrote the best-selling books *The Physics of Star Trek* and *Beyond Star Trek*. In this article, Krauss discusses how the Huygens probe landing on Titan has sparked his imagination and his motivation to be a space exploration enthusiast.

Space Probe Makes Science Fiction Wonders of Childhood Real

By Lawrence M. Krauss
The New York Times, **January 25, 2005**

A small probe stranded on a far-away and hostile world operates for two precious hours at a temperature of 300 degrees below zero Fahrenheit, desperately transmitting information to its mother ship before that spacecraft disappears below the horizon, leaving the small explorer alone on the spongy ground of its new alien home, slowly losing power and slated to eternally rest on a frozen moon 750 million miles from Earth.

I could be accused of anthropomorphizing, but the plight of the small Cassini-Huygens probe resting by a hydrocarbon-coated ice and methane plain on Saturn's largest moon, Titan, captured my imagination far more than anything the astronauts in the International Space Station might be doing now.

What really did it for me was the orange sky. It showed with striking clarity that the science fiction wonders that I dreamed of as a child are being revealed by our unmanned space probes in a way that is both more enthralling and informative than anything likely to come from spending all of NASA's funds on a few more astronauts on the Moon, or, eventually, Mars.

I admit to having already been hooked on Internet images like those from Martian Rovers on a planet that looks suspiciously like a smoggy sunset seen from Los Angeles. But until now, the worlds that were stunningly brought to my desktop were closer to what I might see exploring an earthly desert than to those exotic places that had so captured my imagination as a child reading science fiction stories, or looking at artists' renderings of imaginary planetary surfaces.

But there, as I clicked on the Cassini-Huygens probe Web site, the dark pebbles of dirty hydrocarbon-coated ice on the surface of Titan jumped out through an orange glow of an atmosphere unlike anything I had ever seen.

I was instead reminded of old science fiction stories. On the Web I found a recent example of the kind of thing I used to savor. This was an award-winning short story, "Slow Life" by Michael Swanwick, about human explorers seeking life on Titan.

"People talked a lot about the 'murky orange atmosphere' of Titan, but your eyes adjusted. Turn up the gain on your helmet, and the white mountains of ice were dazzling! The methane streams carved cryptic runes into the heights. Then, at the tholin-line, white turned to a rich palette of oranges, reds and yellows."

So the water-ice is dirtier and the surface darker. But the landscape of Titan is eerily similar to the one Mr. Swanwick imagined so vividly. Except that the truth is even stranger and more entrancing than his fiction.

I learned from a news conference carried out on Friday by the Cassini-Hugyens probe science team that there is evidence of active volcanoes on Titan's surface based on argon 40 in the atmosphere. But these do not spew molten lava. Instead, like the ones I concocted with my childhood chemistry set, these release flumes of water and ammonia.

There are indeed clouds and methane and hydrocarbon rainstorms, but the reality of a turbulent atmosphere of methane winds was brought home to me in a way that no writing could. With brilliant foresight, the Huygens science includes a microphone on the probe. As it fell through the clouds, beginning about 100 miles above the surface, I could listen as well as see the approaching surface as the craft sent out a stream of photos during its descent. Sitting at my computer in the middle of the night, listening to gusts of alien winds on a remote moon of Saturn was both eerie and moving.

I consider myself fortunate to be living at a time when humans are as close as they may ever come to seeing such a truly alien world with methane slush and new colors in the sky. That is probably what drew me to science in the first place. While literature has the power to lift us from the tedium of everyday existence, science at its best has the power to transport us to totally different worlds, both literal and metaphorical, to take us where our imaginations may never have otherwise traveled.

In two short hours, one small unmanned probe changed my direct experience of our solar system in ways that I never imagined. Now I am craving for more such highs. Perhaps I will witness further probes that may dive into distant alien seas underneath frozen moons. Perhaps one will send home clear evidence of alien life existing or extinct.

Realistically, however, the future is likely to be one of cutbacks and shortfalls, with billions of dollars headed to protect populations that we put in jeopardy, or build costly missile defenses against nonexistent threats. One can only hope that there is enough imagination left in government to allow us to keep supporting the missions that do the science that can really change the way we think about our place in the universe.

To boldly go where no one has gone before in ways that only unmanned

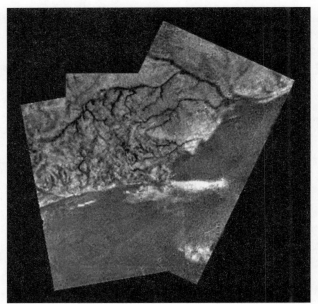

Photo by ESA via Reuters

Images taken by the Cassini-Huygens space probe show details from Saturn's largest moon, Titan, including dark pebbles of dirty hydrocarbon-coated ice.

spacecraft can do will cost so little in comparison that such an effort shouldn't interfere with the current priority of allowing astronauts to have new adventures on the Moon.

It is significant in this regard that the Huygens probe was a product of the European Space Agency, working in concert with NASA and the Jet Propulsion Laboratory. This not only demonstrates that Europe is now a leading player in space exploration, but it shows that for grand human projects, like the exploration of our universe or the exploration of space and time on fundamental scales, we can and need to work together on a global scale.

This is one of the side benefits of the scientific enterprise. But even more than this, the universe continues to surprise us in ways we can never anticipate. Ultimately it is far more interesting than anything that science fiction writers or artists may imagine. Life may imitate art, but ultimately it transcends it. Which is why we sometimes need to turn to the universe itself for inspiration.

IN REVIEW

1. Why would the Huygens probe capture Lawrence Krauss's imagination more than would astronauts on the International Space Station?

2. List the features of Titan that are unlike anything that Krauss has seen before. Explain how these features are linked to his memories of science-fiction stories.

3. People often say that a picture is worth a thousand words. How have the images of Titan's surface and the sounds from the probe's microphone told a story to Lawrence Krauss?

4. Sending robotic missions to planets and moons in the solar system is much more expensive than sending human explorers. What are the benefits and drawbacks of sending humans to interesting places in the solar system compared to sending robots?

5. How has international cooperation enabled the exploration of Titan? In your opinion, should the U.S. continue to partner with foreign countries in space exploration? Explain.

The Cassini/Huygens mission to Saturn has produced many remarkable findings, but perhaps none has been so surprising as the discovery of fountains of ice crystals spraying out from Saturn's moon Enceladus. These ice fountains may imply the existence of an internal liquid zone, which raises at least a slim possibility of life on Enceladus.

Saturn Moon Has Geysers, Hinting Life Is a Possibility

By Kenneth Chang
The New York Times, **March 10, 2006**

With newly discovered signs of liquid water, a moon of Saturn joins the small, highly select group of places in the solar system that could plausibly support life.

The moon, Enceladus, is only 300 miles wide, and usually something that small is nothing more than a frozen chunk of ice and rock. Instead, NASA's Cassini spacecraft has spotted eruptions of icy crystals, which hint at pockets of liquid water near the surface.

"It's startling," said Carolyn C. Porco of the Space Science Institute in Boulder, Colo., leader of the imaging team for Cassini. Nine scientific papers about Enceladus (pronounced en-SELL-ah-dus), appear in today's issue of Science. "I wouldn't be surprised to see the planetary community clamoring for a future exploratory expedition to land on the south polar terrain of Enceladus," said Dr. Porco, lead author of one of the papers. "We have found an environment that is potentially suitable for living organisms."

Life requires at least three ingredients — water, heat and carbon-based molecules — and Enceladus may possess all three. As Cassini flew through the plumes of vapor and ice crystals rising into space from the eruptions, it also detected simple carbon-based molecules like methane and carbon dioxide, which suggest more complicated carbon molecules may lie on the moon's surface.

The lack of a crater suggests that the heat is not the result of a meteor's impact. Based on initial observations, some scientists think that this warm region near the south pole may have persisted for billions of years, sufficient time for life to arise.

"It's an exciting place," said James W. Head III, a professor of geological sciences at Brown University, who was not involved with any of the research reported in today's Science. "That's what exploration is all about. You go out there. It isn't A. It isn't B. It isn't C. It's D, none of the above."

Planetary scientists pointed to the discovery as an argument for continuing NASA's space science efforts. The agency's proposed budget would cut $3 billion from space science over the next five years to help pay for the completion of the space station and plans to send astronauts back to Earth's Moon. NASA's astrobiology institute, which finances research on the possibility of life elsewhere in the solar system and universe, is to see its budget cut in half.

"They must provide sufficient funds for NASA to conduct both human flight and robotic exploration missions," Dr. Porco said. "Right now, the funding is inadequate."

Cassini flew by Enceladus three times last year. For the first two flybys, Cassini's observations of Enceladus' equatorial region turned up nothing odd — except that it seemed to be deflecting Saturn's magnetic fields, implying that Enceladus had a thin atmosphere of charged atoms.

NASA tweaked the trajectory of Cassini's July flyby to pass within about 110 miles of Enceladus' surface. For the first time, the spacecraft got a look at the south pole, which turned out to be smooth compared with the pockmarked northern hemisphere. And it was warm.

The expectation was that the temperature would be about minus-330 degrees Fahrenheit. It turned out more than 100 degrees warmer. "Which is fairly dramatic and blew us away when we first saw it," said John R. Spencer, a planetary scientist at the Southwest Research Institute in Boulder and a team member working with a Cassini instrument that measures infrared emissions. "It's a lot of heat to come out of such a tiny object."

Images of the moon also showed towering plumes of ice crystals coming off at high speed from the surface. The jets seem to originate from fissures near the south pole.

Dr. Porco said calculations eliminated the possibility that the particles were produced by warm vapor rising off warm ice at the surface. The best explanation, she said, is that pockets of liquid water exist under high pressure below a few tens of yards of ice. When the ice ruptures, the water shoots out and immediately freezes into ice crystals.

"We think we've got geysers," Dr. Porco said.

A small body like Enceladus would be unlikely to hold enough radioactive elements to produce continuing warmth. A more likely explanation is that the gravitational tugging on Enceladus by Saturn and another moon, Dione, squishes Enceladus and that friction creates the heat. Another mystery is why the heat is concentrated around the south pole.

IN REVIEW

1. Why was it so startling to find geological activity on a moon the size of Enceladus? How might the necessary heat be generated?

2. What are the characteristics of the surface and of the ice fountains that suggest the existence of a subsurface liquid zone?

3. What are the three basic ingredients of life, and why do we think Enceladus may have all three?

4. Part of the article deals with budgetary concerns of the scientists engaged in the Cassini studies. Summarize their concerns. Overall, do you agree with their sentiments? Defend your opinion.

For first-time telescope users, the target of choice is often Saturn. With its magnificent rings, Saturn is unlike anything that we can see with the naked eye. Larger earth-bound telescopes reveal several distinctive rings, and with the Voyager spacecraft scientists discovered a complicated system of rings and moons in an intricate dance. As scientists looked at Voyager's images they began to create many theoretical models of how Saturn's ring system works, but without the data to support their models, great uncertainties remained. With the Cassini spacecraft now orbiting Saturn, scientists have a platform with several sophisticated scientific instruments that will help untangle the mysteries of Saturn and validate their theoretical ring models.

Decoding the Dance of Saturn's Rings

By Kenneth Chang
The New York Times, July 6, 2004

From far away, Saturn's rings look like a single rigid disk. With more powerful telescopes, 17th-century astronomers were able to discern not a disk but a series of concentric rings.

Even closer, as when NASA's Cassini spacecraft rocketed into Saturn's orbit last week, it becomes clear that the rings are not rigid, either. While Cassini is not sharp-eyed enough to pick out the individual specks and boulders that form the rings, its photographs showed waves disturbing the rings, like ripples radiating from a pebble dropped in a pond. Here, the "pebbles" producing the ripples are Saturn's moons.

For one type of ripple, called density waves, the gravity of a moon outside the ring pulls particles outward. Certain parts of the ring, said to be in resonance with the moon, receive repeated and reinforcing outward kicks.

"Like pushing a kid on a swing," said Dr. Larry W. Esposito, a professor of astrophysical and planetary sciences at the University of Colorado and a member of the Cassini science team. "That changes this uniform density of particles in the ring. The particles clump up."

The dense clump has stronger gravity and draws other particles toward it. That results in a wave slowly spreading outward through the ring. "At about a walking pace," Dr. Esposito said.

The photographs that show the density waves were taken from the back side of the rings, the side facing away from the Sun. The dark bands are not empty regions but rather the denser clumps, which block sunlight. The light bands are the sparser regions where sunlight can pass through.

Because the orbits of many of Saturn's moons are tilted to the rings, the moons' gravity also pulls the rings up and down, creating a second type of ripples called bending waves. Just as shaking out a bedsheet creates waves fluttering down the sheet, the up and down forces create spiraling warps in the rings.

The density waves and the bending waves were not a surprise. Voyager 2 first observed them in August 1981.

Photo by NASA/Jet Propulsion Lab
Cassini postcards: moon-caused ripples and a mysterious striated ring edge.

Rather, the surprise is how well Cassini's photographs appear to fit the theoretical models that scientists have created in the last two decades.

But not everything. A portion of at least one ring appeared less like orderly waves but rather like disheveled straw, probably reflecting a turbulent flow of particles. "Don't have the foggiest," Dr. Esposito said. "Never seen anything like that before."

IN REVIEW

1. Explain how the appearance of Saturn's rings differs when viewed with the naked eye, with small telescopes, with large telescopes, from flyby spacecraft, and from orbiting spacecraft.

2. Describe the types of waves found in Saturn's rings. What causes these waves?

3. Explain how the Cassini spacecraft was positioned to view the waves. What is the importance of position in viewing the rings, and how does a spacecraft provide a unique look?

4. Describe how this article illustrates the hallmarks of science in action. What was confirmed by the data collected by the spacecraft, and what questions remain unanswered?

By now you've surely heard the news: Pluto has been demoted from the list of planets, leaving our solar system with only eight official planets. Under definitions approved in August 2006 by the International Astronomical Union (IAU)—an organization made up of professional astronomers from around the world—Pluto is now considered a "dwarf planet," along with "Xena" (now named Eris) and the asteroid Ceres. Other dwarf planets may be added to the list in the future, since any object that orbits the Sun and is large enough for its own gravity to make it round can qualify. This article, written the day after the official IAU vote, describes how Pluto got its demotion. As background, you should refer to your textbook to remind yourself of how Pluto differs from the first eight planets and of the nature of the *Kuiper belt*.
Note: On September 14, 2006, "Xena" was given the official name Eris; its moon, formerly nicknamed "Gabrielle," is now called Dysnomia.

Vote Makes It Official: Pluto Isn't What It Used to Be

By Dennis Overbye
The New York Times, August 25, 2006

Pluto got its walking papers yesterday. Throw away the place mats. Redraw the classroom charts. Take a pair of scissors to the solar system mobile.

After years of wrangling and a week of debate, astronomers voted for a sweeping reclassification of the solar system. In what many of them described as a triumph of science over sentiment, Pluto was demoted to the status of a "dwarf planet."

In the new solar system as defined by the International Astronomical Union, meeting in Prague, there are eight planets instead of nine, at least three dwarf planets and tens of thousands of so-called smaller solar system bodies, like comets and most asteroids.

For now, the other dwarf planets are Ceres, the largest asteroid, and an object known as 2003 UB 313, nicknamed Xena, that is larger than Pluto and, like it, orbits beyond Neptune in a zone of icy debris known as the Kuiper Belt. But there are dozens more potential dwarf planets known in that zone, planetary scientists say, and so the number in the category could quickly swell.

In a nod to Pluto's fans, the astronomers declared it to be the prototype for a new category of such "trans-Neptunian" objects, but declined in a close vote to approve the name "plutonians" for them.

The outcome yesterday completed a stunning turnaround from only a week ago, when the assembled astronomers were presented a proposal that would have increased the number of planets in the solar system to 12, retaining Pluto and adding Ceres, Xena and even Pluto's moon Charon.

The reversal, said Dr. Alan P. Boss, a planetary theorist at the Carnegie Institution of Washington, speaks to the integrity of the planet defining process.

"The officers were willing to change their resolution," Dr. Boss said, "and find something that would stand up under the highest scientific scrutiny and be approved."

Jay M. Pasachoff, a Williams College astronomer who attended the Prague congress and favored somehow keeping Pluto a planet, said, "The spirit of the meeting was of future discovery and activity in science rather than any respect for the past."

Mike Brown of the California Institute of Technology, who discovered UB 313 three years ago and so had the most to lose personally from the downgrading of Pluto and Xena, said he was relieved.

"Through this whole crazy circus-like procedure, somehow the right answer was stumbled on," Dr. Brown said. "It's been a long time coming. Sci-

ence is self-correcting eventually, even when strong emotions are involved."

It had long been clear that Pluto, discovered in 1930, stood apart from the previously discovered planets. Not only is it much smaller — only about 1,600 miles in diameter, smaller than the Moon — but its elongated orbit is tilted with respect to the other planets, and it goes inside the orbit of Neptune on part of its 248-year journey around the Sun.

Pluto, some astronomers had argued, made a better match with the other ice balls that have since been discovered in the dark realms beyond Neptune. In 2000, when the Rose Center for Earth and Space opened at the American Museum of Natural History, Pluto was denoted in a display as a Kuiper Belt object and not a planet.

In the decision yesterday as to what constitutes a planet, astronomers voted by standing and holding up yellow cards. In the crucial vote, the result was sufficiently one-sided that no formal count was taken.

Under the new rules, a planet must meet three criteria: it must orbit the Sun, it must be big enough for gravity to squash it into a round ball, and it must have cleared other things out of the way in its orbital neighborhood. The last of these criteria knocks out

Pluto and Xena, which orbit among the icy wrecks of the Kuiper Belt, and Ceres, which is in the asteroid belt.

Dwarf planets, on the other hand, need only orbit the Sun and be round.

"I think this is something we can all get used to as we find more Pluto-like objects in the outer solar system," Dr. Pasachoff said.

The final voting was by some 400 to 500 of the 2,400 astronomers who registered for the congress; many others had already left.

Pointing to the very small fraction of the world's astronomers who had been in Prague and thus eligible to vote, Alan Stern, lead investigator for New Horizons, NASA's mission to Pluto, called the resolution "laughable." Dr. Stern, of the Southwest Research Institute in Boulder, Colo., pointed out that both Earth and Jupiter have asteroids in their neighborhoods.

"This is so scientifically sloppy and internally inconsistent," he wrote in an e-mail message, "that it is embarrassing."

This is not the first time that astronomers have rethought a planet. The asteroid Ceres was hailed as the eighth planet when it was discovered in 1801 by Giovanni Piazzi, floating between Mars and Jupiter. But the subsequent discovery of more and more things like it in the same part of space led astronomers to dub them asteroids.

Although many astronomers watched the vote on the Internet, Neil deGrasse Tyson of the Rose Center said he had not bothered.

"Counting planets is not an interesting exercise to me," Dr. Tyson said. "I'm happy however they choose to define it. It doesn't really make any difference to me."

Far more compelling, he added, are aspects of planets like weather, ring systems and magnetic fields.

Dr. Tyson said a continuing preoccupation with what the public and schoolchildren would think about this was a concern and a troubling precedent.

"I don't know any other science that says about its frontier, 'I wonder what the public thinks,' " he said. "The frontier should move in whatever way it needs to move."

IN REVIEW

1. Why has the planetary status of Pluto long been controversial?

2. Under the new definitions, why isn't Pluto a true planet?

3. Under the new definitions, what is a *dwarf planet*? What solar system objects qualify?

4. Based on this article, briefly describe how the decision was made, and some of the reactions of other astronomers to it. Of the astronomers quoted, whose reaction do you most agree with? Why?

5. As the article makes clear, at least some astronomers are already objecting to the new definitions. For further background, read the op-ed piece below by Jeffrey Bennett (the lead author of your textbook). After reading both the *New York Times* article and Bennett's article, write a short opinion piece of your own in which you describe how you personally feel about the new definitions, and whether you think the definitions will stick over time. Defend your opinions.

Bulldoze Pluto? I Don't Think So
Jeffrey Bennett
8/25/06
www.jeffreybennett.com

The International Astronomical Union (IAU) has spoken on the status of Pluto. The only thing missing when they announced the decision at their press conference was the "Mission Accomplished" banners. Yes, I'm afraid this matter is far from settled.

You've probably heard the basics: Pluto is no longer to be considered a "real" planet, but will instead be part of a new class of objects called "dwarf planets." These midgets may number anywhere between a handful and hundreds in our own solar system, depending on how you count them and how hard we search for them with more powerful telescopes. But no matter how you cut it, this new definition takes away any pretense of Pluto being a member of the same elite, planetary club as Earth and Jupiter.

As a scientist, I think this was a pretty good outcome, though some of the justifications used to achieve it are dubious. But as a teacher, textbook writer, and builder of scale model solar systems, I have some reservations. In particular, for the model solar systems I've helped develop on the University of Colorado Boulder campus and on the National Mall in Washington, DC, I have to ask: Should we now bulldoze Pluto?

A little background: Pluto was discovered in 1930, at a time when astronomers were searching for an object thought to be causing slight perturbations to the orbits of other planets around the Sun. Neptune itself had been discovered in just this way in 1846, after scientists used perturbations in Uranus's orbit to predict the existence and location of an "eighth planet." Neptune was found as soon as astronomers pointed telescopes to the calculated position, which is I like to say that Neptune was discovered with physics and mathematics, and only confirmed with a telescope.

Over the ensuing decades, a few scientists claimed to see ongoing orbital discrepancies and embarked on a search for a "ninth planet" that might be causing them. Pluto was found during this search, though about 12 full-moon-widths away from the predicted position. And though hailed as a planet upon discovery, its status gradually became suspect, as we learned that its orbit is much more tilted and elongated than that of any of the other planets, and that it has a mass much less than 1% that of Earth. Worse, reanalysis of past observations suggested that the claimed orbital discrepancies had simply been errors in measurement, making Pluto a solution to a non-existent problem.

In 1951, by analyzing comet orbits, astronomer Gerard Kuiper predicted the existence of a "comet belt" beyond Neptune — now officially named the *Kuiper belt* — analogous to the asteroid belt between Mars and Jupiter. Telescope technology caught up with this prediction in the 1990s, and astronomers soon confirmed the existence of vast numbers of "Kuiper belt objects" orbiting the Sun in the general vicinity of Pluto. Scientifically, it became obvious that Pluto was much more like a large member of this group than like a small version of any of the other 8 planets. But as long as Pluto was the *largest* Kuiper belt object, the IAU felt it acceptable to leave its planetary status intact.

Then, in July 2005, astronomer Mike Brown announced the discovery of a new Kuiper belt object — still nameless today but nicknamed "Xena" or "Planet X" — that is slightly larger than Pluto. This discovery was a fatal blow to the status quo of nine planets; after all, if Pluto is big enough to count as a planet, then Xena must belong to the club too. But what of the dozens of Kuiper belt objects only a little smaller than Pluto, and of future discoveries yet to be made? How would we draw the line between "planet" and "large Kuiper belt object?" Like all groups facing a tough decision, the IAU appointed a committee.

At first, the committee proposed roundness as a primary criterion for planethood, which would have admitted both Pluto and Xena to the club. However, their proposal would have also admitted Pluto's moon Charon (for technical reasons), perhaps as many as 40 or more other Kuiper belt objects, and even the asteroid Ceres. Given that Ceres was discovered in 1801, we would have had to start saying silly things like "The ninth planet, Neptune, was discovered in 1846, when it was mistakenly identified as the eighth planet."

The IAU members saw the light, and modified the definition to keep Ceres and Charon out of the club by adding that a planet must not only be round, but must also have "cleared the neighborhood around its orbit." Ceres fails this criterion, because it is just one of many asteroids in the asteroid belt. Charon returns to being a moon of Pluto. But Pluto loses its status too, because it is just one of many objects in the Kuiper belt, and in fact crosses over the orbital path of Neptune.

Some of my colleagues are quite upset, pointing out that no object has truly cleared its orbit — that's why collisions still sometimes occur, like the comet that smacked into Jupiter in 1994 or the asteroid that struck Siberia in 1908. Others wonder why we need to add the new term "dwarf planet," when "large asteroid" or "large Kuiper belt object" can already describe objects like Ceres and Pluto and Xena quite well. And imagine the confusion if we someday discover a Mars-or Earth-size iceball in the Kuiper belt: officially, such an object would count as a dwarf planet, but it would be larger than some of the non-dwarfs. Indeed, many astronomers wonder why we need an official definition of "planet" at all. That's why I doubt this debate is over, and it brings me back to my point: do we really need to take Pluto out of songs, place mats, mobiles, and model solar systems? I don't think so.

The important thing to remember is that the new definition was established by a vote, making it politics, not science. The politics may yet change again, but Pluto will remain the same. I suspect it will stay a planet in hearts and minds no matter what the IAU says, much as Europe and Asia remain separate continents to everyone except the geologists. Pluto, after all, is not just *any* planet — it's the *only* planet with a famous dog named after it, and some people may not take well to sending the dog home just because it didn't turn out quite as big as originally hoped. So let's keep the bulldozers away from the Pluto pedestals on the CU campus and the National Mall. In fact, if someone will provide the budget, I'd recommend adding a pedestal for Xena, if astronomers ever vote to give it a real name.

As discussed in your text, Pluto is just one of more than 1,000 known objects orbiting the Sun in the region beyond Neptune known as the Kuiper belt. At least one of these objects (nicknamed "Xena" in the article but now named Eris) is larger than Pluto, and many more may be relatively close in size to Pluto. This article discusses some of the characteristics of these many Kuiper belt objects, along with some new ideas about solar system formation based on the study of these icy objects.
Note: On September 14, 2006, "Xena" was given the official name Eris; its moon, formerly nicknamed "Gabrielle," is now called Dysnomia.

Pluto's Exotic Playmates

By Kenneth Chang
The New York Times, **September 12, 2006**

With a quick vote last month, the International Astronomical Union decreed that Pluto was no longer the ninth planet, but just a dwarf planet — and not even the largest dwarf — orbiting in a distant ring of icy debris.

But perhaps that should not be seen as a slight to Pluto.

For many astronomers, that ring of icy debris, known as the Kuiper Belt, has become an exciting spot for innovative research and has changed how they view the solar system.

"It's a lot bigger now," said Marc W. Buie, an astronomer at the Lowell Observatory in Flagstaff, Ariz. "For me, it's like somebody invented a new field of science."

Harold F. Levison of the space studies department in the Southwest Research Institute in Boulder, Colo., said, "The more we learn, the weirder it looks."

More than 1,100 Kuiper Belt objects have been found so far. Astronomers estimate that half a million bodies larger than 20 miles wide are floating out there. At least one appears to be mostly rock with a coating of ice. Some are mostly ice. Some are less dense than ice, indicating a Swiss-cheese-like structure. A surprising number of them have moons.

Some move in clockwork with Neptune; Pluto, for example, is in what is called a 3:2 resonance, taking 1.5 times as long as Neptune to loop the Sun. Many Kuiper Belt objects have been flung into orbits crazily tilted to the rest of the solar system.

"This is really a very exotic zoo out there," said S. Alan Stern, executive director of the space science and engineering division at the Southwest Research Institute and principal investigator of NASA's New Horizon spacecraft, which is currently heading to Pluto.

The distribution of Kuiper Belt objects has already provided decisive evidence that Neptune was once perhaps nearly a billion miles closer to the Sun and was then gravitationally nudged outward. Astronomers also hope that the Kuiper Belt preserves a frozen record of the earliest building materials of the solar system.

"It's kind of like the solar system's attic," Dr. Stern said. "It's like an archaeological dig into the history of our solar system."

Scientists had initially expected a simple structure for the belt: a thin disk of objects traveling in circular orbits in the plane of the solar system. Some Kuiper Belt objects do fit that profile, and those are called the classical Kuiper Belt objects. (One mystery is why there appears to be a sharp edge at about 4.5 billion miles, with no classical Kuiper Belt objects beyond that distance. Some think a passing star did that.)

Other Kuiper Belt objects share orbits similar to Pluto's, in resonance with Neptune. Those in the same 3:2 resonance as Pluto have been called the Plutinos.

Still others are called the scattered-disk Kuiper Belt objects. These appear to have been tossed into highly elliptical orbits, often at a sharp angle to the rest of the solar system. Surprisingly, these include some of the larger Kuiper Belt objects, including 2003 UB313, nicknamed Xena, which is larger than Pluto.

An estimated 15 percent of Kuiper Belt objects are binaries — pairs of bodies of similar size and mass. Among some classical Kuiper Belt objects, that fraction may be as high as 30 percent — possibly higher, because even the Hubble Space Telescope cannot distinguish two separate objects if they are too close to each other.

Theorists puzzled about how such small bodies, with weak gravitational pull, could have paired up so often. The answer, it turns out, is that as two objects flew past each other, the gravitational drag generated by many other much smaller Kuiper Belt objects slowed them enough to capture each other.

That mechanism requires a fairly dense Kuiper Belt with a total mass of at least 10 Earths. But while Kuiper Belt objects are many, they do not amount to much today. Adding the masses of Pluto, Xena and the half million other objects, even those not yet seen, gives an estimate of just one-tenth the mass of Earth.

"The mass we measure is pathetic," said David C. Jewitt, a professor of astronomy at the University of Hawaii.

That, in turn, produces a quandary. Where has 99 percent of the Kuiper Belt gone? This is, as the planetary scientists quaintly put it, the cleanup problem. (One popular idea: repeated collisions smashed most of them to dusty bits, and the bits were blown away by solar radiation.)

Just 15 years ago, the Kuiper Belt was not on the map at all. The known solar system essentially ended at Neptune, except for the occasional comet interloper from far away. (Fifteen years ago, Pluto was traveling along the inner part of its eccentric orbit, closer to the Sun than Neptune.)

Gerard Kuiper, a prominent astronomer for whom the belt is named, speculated about the possibility of it in 1951. Kuiper was prescient, but wrong on a major point. He thought that the belt had existed early in the solar system's history but that Pluto, then thought to be a more massive planet, had scattered it away.

Some people have suggested that the belt should instead be named after Kenneth Edgeworth, an Irish astronomer who vaguely hypothesized an icy disk a few years before Kuiper. But that was just a guess.

"The outer solar system was just this sort of empty space," said Michael E. Brown, a professor of planetary astronomy at the California Institute of Technology.

Astronomers did have a few clues that something was out there. One was Pluto, an oddball. Unlike the four rocky inner planets or the four outer gas giants, Pluto is half ice, and its orbit is quite elliptical and tilted 17 degrees to the solar system's ecliptic plane. It did not fit in.

A second clue showed up in 1977 with the discovery of Chiron, an icy body between 90 and 130 miles wide that loops around on an elliptical path taking it as close to the Sun as Saturn and as far out as Uranus. Astronomers argued over whether it should be called an asteroid or a comet. In the end, they decided both, and created a new category, Centaurs, to describe small bodies orbiting among the giant planets.

More interestingly, Chiron's orbit is unstable, meaning that within a million years or so, it will probably swing too close to Saturn and be tossed out of the solar system or onto a cometlike trajectory passing closer to the Sun. That also means that Chiron entered its current orbit in astronomically recent times. Like a lamb wandering the streets of Manhattan, it had to have come from somewhere else. No one knew where.

A third, crucial clue came from comets. Some comets, known as long-period comets, visit the inner solar system once in thousands or millions of years, or simply once. Others like Halley's Comet are short-period comets that swing by more frequently, every few decades or centuries or so.

For a long time, most astronomers theorized that short-period comets were long-period comets that had been deflected by the gravity of a planet.

In 1988, computer simulations by three Canadian astrophysicists — Martin J. Duncan, Thomas R. Quinn and Scott Tremaine — showed that the orbits of deflected long-period comets would not match those of observed short-period comets and that their more likely origin was a ring of debris by Neptune. Those computer simulations spurred several teams to start searching more actively the outskirts of Neptune.

Dr. Jewitt, then at the Massachusetts Institute of Technology, and his graduate student Jane X. Luu had already started looking.

"Our search was very simply motivated by the surprising emptiness at the edge of the solar system," Dr. Jewitt said. "We would have been happy with either answer: empty because it was empty, or empty because no one had looked."

For years, they found nothing. To find moving objects requires taking repeated photographs of a region and looking for the points of light that moved between the images. The limits of digital technology stymied the early searches.

Finally, in 1992, they found 1992 QB1, probably about 100 miles wide, the first Kuiper Belt object.

Dr. Brown had a reaction similar to many astronomers. "It's like, 'Oh my god, Pluto finally makes sense,'" he said. "It's no longer an oddball at the edge of the solar system."

Pluto was instead the harbinger of many properties now seen in Kuiper Belt objects: the resonant orbits, the moons, the icy ingredients.

About that time, Renu Malhotra, a professor of astronomy at the University of Arizona, was calculating the effects of Neptune's migrating outward, as some had hypothesized, early in the history of the solar system. Her calculations indicated that Neptune would effectively snowplow smaller objects into resonant orbits.

The first Kuiper Belt objects in resonant orbits were discovered in 1993. The distribution of resonant Kuiper Belt objects fit with what she predicted, and now planetary scientists uniformly agree that the giant planets were not born in their current orbits but migrated there.

Scientists are looking for more clues about the early history of the solar system. Eugene Chiang, a professor of astronomy at the University of California, Berkeley, theorizes that the solar system used to contain several more gas giant planets that were subsequently ejected, but that their gravitational effects remain imprinted in the Kuiper Belt.

Dr. Levison of the Southwest Research Institute, however, offers a different account. He and his collaborators have created a model where the solar system was initially much more compact, with all the giant planets forming well within the current orbit of Uranus.

That hypothesis sidesteps the cleanup problem, because shrinking the Kuiper Belt increases its density. His model predicts that gravitational wobbling between Jupiter and Saturn created wild oscillations in the orbits of Neptune and Uranus, with the two swapping places repeatedly.

Dr. Levison said he could not prove that his model was correct, just that it reproduces what is seen today in the solar system. And the Kuiper Belt was a key component in creating his model.

"Sometimes how the blood is splattered on the wall tells you more about what happened than the body," he said.

The Kuiper Belt, Dr. Levison said, is "the blood splattering on the wall," adding, "If we're going to understand what happened, it's going to be by studying the Kuiper Belt."

IN REVIEW

1. What types of orbital resonances are seen among the Kuiper belt objects? What ideas can explain these resonant orbits?

2. How many of the Kuiper belt objects are binaries, and why is this surprising?

3. How did the discovery of Kuiper belt objects help astronomers "make sense" of Pluto?

4. How do the orbital characteristics of Kuiper belt objects lend support to the idea that the jovian planets migrated outward in the early history of the solar system?

All known extrasolar planets are considerably larger than Earth. However, as discussed in your text, this fact almost certainly reflects only technological limitations, in that larger planets are much easier to detect than smaller ones. Indeed, as technology improves, the "record" for the smallest known extrasolar planet has been falling rapidly. This article describes the discovery of a planet that, if real, is the smallest yet discovered as of 2006.

Search Finds Far-Off Planet Akin to Earth

By Dennis Overbye
The New York Times, **January 26, 2006**

Now you see it, now you don't. Astronomers say that by virtue of the ceaseless shifting of the billions of stars in the Milky Way and a trick of Einsteinian physics, they have briefly glimpsed the most Earth-like planet yet to be discovered outside the solar system. It is a ball of rock and ice only about 5.5 times as massive as Earth, smaller than any of the 160 previously discovered exoplanets, and is orbiting a dim reddish star 21,000 light-years from here.

The discovery, the researchers say, suggests that rock-ice planets like our own are predominant in the cosmos. That bodes well for future planet-hunting missions from space like the Terrestrial Planet Finders at NASA.

The distant planet manifested itself as a brief flash: As it passed at night in front of an even more distant star, its gravity focused and momentarily brightened the star's light.

It was all over in less than a day, a cosmic blink of an eye.

"It was the blip in the night that we have been waiting for," said Jean-Philippe Beaulieu of the Institut d'Astrophysique de Paris, who led an international collaboration of 73 astronomers. They reported their findings yesterday at a news conference in Washington and are publishing them today in the journal Nature.

Alan P. Boss, a planetary theorist at the Carnegie Institution of Washington, said in an e-mail message that the discovery was "a big one." Another expert who took no part in the research, Geoffrey W. Marcy of the University of California, Berkeley, said, "The result looks solid to me, and perhaps the planet is, too."

The planet, smaller than Neptune and dubbed OGLE-2005-BLG-390Lb, resides about 234 million miles from its star. At that distance, its surface temperature would be minus 370 degrees Fahrenheit, Dr. Beaulieu said.

The work was largely that of two large teams that have built far-flung observing networks to exploit a feature of Einstein's general theory of relativity. The theory says a massive object can act as a gravitational lens, bending and magnifying the light from more distant objects in space.

One team, the Optical Gravitational Lensing Experiment, or OGLE, led by Andrzej Udalski of Warsaw University, has been set up to monitor the brightness of millions of Milky Way stars every night from a telescope in Chile, so as to catch fluctuations caused by the passage of intervening objects of various kinds: dim stars, so-called dark matter objects or planets.

Last July, alerted by the OGLE team that such a lensing event was under way, Dr. Beaulieu's team, Planet (for Probing Lensing Anomalies NETwork), sprang into action to do high-resolution observations. On Aug. 9, that team recorded a tiny blip on a much larger blip that was caused by the passage of an unseen star, the planet's parent, in front of the more distant star. That tiny blip was the planet itself.

As Dr. Beaulieu explained in an e-mail message, it would have been much easier to see a giant gaseous type of planet. The long odds against detecting so small a planet as the new one argues for its commonness.

"If only a small fraction of the stars had such planets, we would have never detected this small planet," Dr. Beaulieu said.

Scott Tremaine, a theorist at Princeton, said, "The results suggest that rock-ice planets must be more common than gas giants."

IN REVIEW

1. What is OGLE, and how does it work?

2. How did the OGLE observations lead to the discovery of this new planet?

3. What is the estimated mass of the new planet? What is its estimated distance from its star? Given these estimates, briefly describe the likely characteristics of the planet.

4. One drawback to the OGLE observations is that because they detect passages in front of more distant stars—as opposed to, say, repeated orbits of a planet around its star — the observed events occur only once. This means, for example, that we have no hope of repeating the observation, or of learning the orbital period of the planet. Should this affect how seriously we take the result? Defend your opinion.

Although there is debate about whether Pluto should count as a planet, one fact is indisputable: Among the nine planets known prior to 2005, Pluto is the only one never visited by a spacecraft. The *New Horizons* mission will change that, as it speeds toward a rendezvous with this enigmatic world.

NASA Launches Spacecraft On the First Mission to Pluto

By Warren E. Leary
The New York Times, **January 20, 2006**

NASA launched the first space mission to Pluto yesterday as a powerful rocket hurled the New Horizons spacecraft on a nine-year, three-billion-mile journey to the edge of the solar system.

As it soared toward a 2007 rendezvous with Jupiter, whose powerful gravitational field will slingshot it on its way to Pluto, mission managers said radio communications confirmed that the 1,054-pound craft was in good health.

The $700 million mission began when a Lockheed Martin Atlas 5 rocket rose from a launching pad at the Cape Canaveral Air Force Station in Florida at 2 p.m., almost an hour later than planned because of low clouds that obscured a clear view of the flight path by tracking cameras.

"We have ignition and liftoff of NASA's New Horizons spacecraft on a decadelong voyage to visit the planet Pluto and then beyond," declared Bruce Buckingham, NASA's launching commentator.

Less than an hour later, all three stages of the booster rocket worked as planned, and the spacecraft separated from them and sprinted away toward deep space. The robot ship sped away at about 36,000 miles per hour, the fastest flight of any spacecraft sent from Earth, allowing it to pass the Moon in about nine hours.

"This is a historic day," said Alan Stern of the Southwest Research Institute in Boulder, Colo., the mission's principal scientist and team leader.

Speaking at a news conference at the Kennedy Space Center in Florida, Dr. Stern said the timing assured that the New Horizons would arrive for its closest approach to Pluto on July 14, 2015 — the 50th anniversary of the first flyby of Mars by the Mariner 4, the mission that began the exploration of the planets.

Yesterday's liftoff also paid homage to Pluto's discoverer, the astronomer Clyde W. Tombaugh, who in 1930 became the only American to find a planet in the solar system. (He died at 90, in 1997.) His widow, Patricia Tombaugh, 93, and other family members were present at the cape, and some of his remains were among the commemorative items aboard the spacecraft.

"Some of Clyde's ashes are on their way to Pluto today," Dr. Stern said.

The New Horizons is to reach Jupiter's gravitational field in 13 months. The trip to Pluto will take eight more years, most of which the craft will spend in electronic "hibernation" to save power and wear on the equipment needed for its seven experiments.

The New Horizons is powered by a small plutonium-fired electric generator. Its instruments include three cameras, for visible-light, infrared and ultraviolet images, and three spectrometers to study the composition and temperatures of Pluto's thin atmosphere and surface features.

It also carries a University of Colorado dust counter, the first experiment to fly on a planetary mission that is

entirely designed and operated by students. This is the only experiment that will not hibernate during the mission.

In addition to the two-hour delay, the launching was postponed twice in two days — on Tuesday by strong winds at the cape and on Wednesday by a storm that caused a power failure at the spacecraft's control center at the Johns Hopkins University Applied Physics Laboratory in Laurel, Md. Mission planners had until Feb. 14 to launch the mission this year, but only until the end of this month to use the gravity boost from Jupiter, which will shorten the trip to Pluto by five years.

Once near its target, the New Horizons is to conduct about five months of studies, including a closest-approach dash that takes it within 6,200 miles of Pluto's surface and 16,800 miles from the planet's large moon, Charon. The craft will also study two smaller moons found late last year by the Hubble Space Telescope and any new features discovered while it is on its way, scientists said.

The mission is to continue past Pluto, possibly visiting large objects in the Kuiper Belt, an outer zone of the solar system that includes Pluto. The belt is made up of thousands of icy, rocky objects that include comets and small planets. Scientists believe that this material is left over from the creation of the solar system 4.6 billion years ago and that studying it will provide clues to how the Sun and planets formed.

IN REVIEW

1. When was *New Horizons* launched? When will it reach Pluto?

2. Will *New Horizons* orbit Pluto or fly by? Why was this option chosen?

3. *New Horizons* is the fastest spacecraft ever launched. How fast was it going? Put this speed in perspective by comparing it with some more familiar speeds or distances that can be covered at its speed.

4. Briefly describe the *New Horizons* spacecraft and the purposes of its scientific instruments. What experiment will operate throughout the mission? Why will others go into "hibernation" for most of the journey?

5. What will the spacecraft do as it approaches Pluto? What do scientists hope to learn?

The movement of the Earth's crust, which is broken into plates, is one of the unique features of our planet. As more than a dozen plates slowly collide and separate, mountain ranges rise, rift valleys split, and volcanic islands emerge. More sudden and violent plate movement results in earthquakes. Such was the case in December 2004, when a release of huge stresses between two crustal plates near Sumatra, Indonesia, created a violent earthquake below the ocean. The resulting tsunami caused a wave of destruction for thousands of miles around. Understanding the motion of the plates and the resulting impacts is the work of countless numbers of scientists who seek, in part, to prevent the huge loss of life associated with catastrophic tsunamis.

Deadly and Yet Necessary, Quakes Renew the Planet

By William J. Broad
The New York Times, **January 11, 2005**

They approach the topic gingerly, wary of sounding callous, aware that the geology they admire has just caused a staggering loss of life. Even so, scientists argue that in the very long view, the global process behind great earthquakes is quite advantageous for life on earth—especially human life.

Powerful jolts like the one that sent killer waves racing across the Indian Ocean on Dec. 26 are inevitable side effects of the constant recycling of planetary crust, which produces a lush, habitable planet. Some experts refer to the regular blows—hundreds a day—as the planet's heartbeat.

The advantages began billions of years ago, when this crustal recycling made the oceans and atmosphere and formed the continents. Today, it builds mountains, enriches soils, regulates the planet's temperature, concentrates gold and other rare metals and maintains the sea's chemical balance.

Plate tectonics (after the Greek word "tekton," or builder) describes the geology. The tragic downside is that waves of quakes and volcanic eruptions along plate boundaries can devastate human populations.

"It's hard to find something uplifting about 150,000 lives being lost," said Dr. Donald J. DePaolo, a geochemist at the University of California, Berkeley. "But the type of geological process that caused the earthquake and the tsunami is an essential characteristic of the earth. As far as we know, it doesn't occur on any other planetary body and has something very directly to do with the fact that the earth is a habitable planet."

Many biologists believe that the process may have even given birth to life itself.

The main benefits of plate tectonics accumulate slowly and globally over the ages. In contrast, its local upheavals can produce regional catastrophes, as the recent Indian Ocean quake made clear.

Even so, scientists say, the Dec. 26 tsunamis may prove to be an ecological boon over the decades for coastal areas hardest hit by the giant waves.

Dr. Jelle Zeilinga de Boer, a geologist at Wesleyan University who grew up in Indonesia and has studied the archipelago, says historical evidence from earlier tsunamis suggests that the huge waves can distribute rich sediments from river systems across coastal plains, making the soil richer.

"It brings fertile soils into the lowlands," he said. "In time, a more fertile jungle will develop."

Dr. de Boer, author of recent books on earthquakes and volcanoes in human history, added that great suffering from tectonic violence was usually followed by great benefits as well. "Nature is reborn with these kinds of terrible events," he said. "There are a lot of positive aspects even when we don't see them."

Plate tectonics holds that the earth's surface is made up of a dozen or so big crustal slabs that float on a sea of melted rock. Over ages, this churning sea moves the plates as well as their superimposed continents and ocean basins, tearing them apart and rearranging them like pieces of a puzzle.

The process starts as volcanic gashes spew hot rock that spreads out across the seabed. Eventually, hundreds or thousands of miles away, the cooling slab collides with other plates and sinks beneath them, plunging back into the hot earth.

The colliding plates grind past one another about as fast as fingernails grow and over time produce mountains and swarms of earthquakes as frictional stresses build and release. Meanwhile, parts of the descending plate melt and rise to form volcanoes on land.

The recent cataclysm began in a similar manner as volcanic gashes in the western depths of the Indian Ocean belched molten rock to form the India plate. Its collision with the Burma plate created the volcanoes of Sumatra as well thousands of earthquakes, including the magnitude 9.0 killer.

A Lush Planet Emerges From Epic Violence

Earthquakes and resulting tsunamis are unfortunate byproducts of a process necessary to renew the earth. Subduction — the collision of plates that make up the earth's crust in which one plate is pulled below another — plays a critical role in recycling carbon dioxide, which is needed to regulate the planet's temperature.

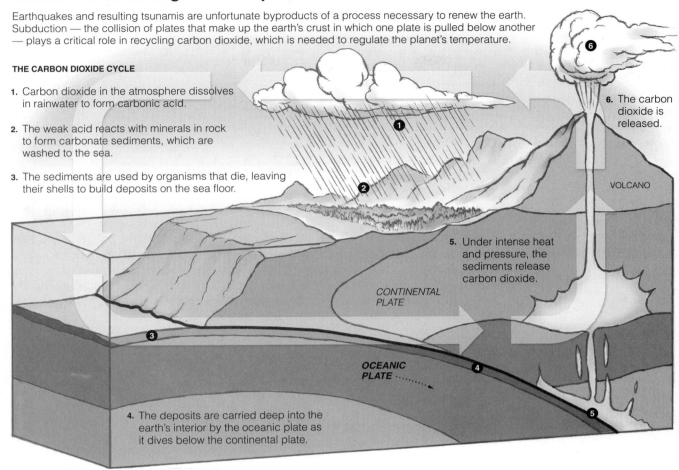

THE CARBON DIOXIDE CYCLE

1. Carbon dioxide in the atmosphere dissolves in rainwater to form carbonic acid.

2. The weak acid reacts with minerals in rock to form carbonate sediments, which are washed to the sea.

3. The sediments are used by organisms that die, leaving their shells to build deposits on the sea floor.

5. Under intense heat and pressure, the sediments release carbon dioxide.

6. The carbon dioxide is released.

VOLCANO

CONTINENTAL PLATE

OCEANIC PLATE

4. The deposits are carried deep into the earth's interior by the oceanic plate as it dives below the continental plate.

Other Benefits of Plate Tectonics

 Made oceans and atmosphere
Most geologists believe the earth's primordial ocean and atmosphere arose as volcanoes spewed water vapor, nitrogen, carbon dioxide and other gases.

 Made land
Simple volcanism made islands like Sumatra and Java, where scores of volcanoes still erupt. Additionally, geologists say plate tectonics builds continents and mountain ranges.

 Improves soil
Volcanic soils, some of the world's richest, are often used to form plantations for producing coffee, sugar, rubber, coconuts, palm oil, tobacco, pepper, tea and cocoa.

 Begat life
Many biologists believe the earth's first organisms arose in the deep sea along volcanic gashes and that microbes known as thermophiles, which thrive today in such hot regions, are their direct descendants.

 Regulates global chemistry
Experts say the world ocean passes through the tectonic system once every million years or so, increasing nutrients in the biosphere and regulating a host of elements and compounds.

 Concentrates metals
Water streaming through the seabed's hot volcanic gashes concentrates silver, gold and other metals into rich deposits that are mined after plate tectonics nudges them up onto land.

Sources: "Biogeochemistry: An Analysis of Global Change" by William H. Schlesinger; "Understanding Earth" by Frank Press, Raymond Siever, John Grotzinger and Thomas H. Jordan

David Constantine, Al Granberg/The New York Times

But despite such staggering losses of life, said Robert S. Detrick Jr., a geophysicist at the Woods Hole Oceanographic Institution, "there's no question that plate tectonics rejuvenates the planet."

Moreover, geologists say, it demonstrates the earth's uniqueness. In the decades after the discovery of plate tectonics, space probes among the 70 or so planets and moons that make up the solar system found that the process existed only on earth—as revealed by its unique mountain ranges.

In the book "Rare Earth" (Copernicus, 2000), which explored the likelihood that advanced civilizations dot the cosmos, Dr. Peter D. Ward and Dr. Donald Brownlee of the University of Washington argued in a long chapter on plate tectonics that the slow recycling of planetary crust was uncommon in the universe yet essential for the evolution of complex life.

"It maintains not just habitability but high habitability," said Dr. Ward, a paleontologist. (Dr. Brownlee is an astronomer.) Most geologists believe that the process yielded the earth's primordial ocean and atmosphere, as volcanoes spewed vast amounts of water vapor, nitrogen, carbon dioxide and other gases. Plants eventually added oxygen. Meanwhile, many biologists say, the earth's first organisms probably arose in the deep sea, along the volcanic gashes.

"On balance, it's possible that life on earth would not have originated without plate tectonics, or the atmosphere, or the oceans," said Dr. Frank Press, the lead author of "Understanding Earth" (Freeman, 2004) and a past president of the National Academy of Sciences.

The volcanoes of the recycling process make rich soil ideal for producing coffee, sugar, rubber, coconuts, palm oil, tobacco, pepper, tea and cocoa. Water streaming through gashes in the seabed concentrates copper, silver, gold and other metals into rich deposits that are often mined after plate tectonics nudges them onto dry land.

Experts say the world ocean passes through the rocky pores of the tectonic system once every million years or so, increasing nutrients in the biosphere and regulating a host of elements and compounds, including boron and calcium.

Dr. William H. Schlesinger, dean of the Nicholas School of the Environment and Earth Sciences at Duke, says one vital cycle keeps adequate amounts of carbon dioxide in the atmosphere. Though carbon dioxide is thought to cause excessive greenhouse-gas warming of the planet, an appreciable level is needed to keep the planet warm enough to support life.

"Having plate tectonics complete the cycle is absolutely essential to maintaining stable climate conditions on earth," Dr. Schlesinger said. "Otherwise, all the carbon dioxide would disappear and the planet would turn into a frozen ball."

Dr. Press, who was President Jimmy Carter's science adviser, said the challenge in the coming decades would be to keep enjoying the benefits of plate tectonics while improving our ability to curb its deadly byproducts.

"We're making progress," Dr. Press said. "We can predict volcanic explosions and erect warning systems for tsunamis. We're beginning to limit the downside effects."

IN REVIEW

1. Identify and list the different effects of plate tectonics.

2. Contrast the gradual effects of plate tectonics with those that happen rapidly.

3. Describe how plate tectonic cycles were demonstrated by the December 2004 earthquake near Sumatra.

4. How would plate tectonics make Earth highly habitable? Compare the major life-supporting benefits of the Earth's plate tectonics to a planet like Mars that does not have plate tectonics. What are the implications for finding life on Mars?

5. How will understanding plate tectonics better prevent major losses of life associated with earthquakes and tsunamis? How would deepening our knowledge through science benefit the technological advances made by engineers and entrepreneurs who are concerned with public safety?

As discussed in your text (and daily in the news media!), global warming induced by the human emission of greenhouse gases is one of the most important issues of our time. Although the current period of global warming is unique in being caused by us, Earth's climate has undergone large climate swings at many times in the past. The more we can learn about the past climate, the better we'll understand the likely consequences of global warming today. This article describes recent findings about the Arctic climate going back more than 55 million years.

Studies Portray Tropical Arctic as Sultry in Distant Past

By Andrew C. Revkin
The New York Times, June 1, 2006

Correction Appended

The first detailed analysis of an extraordinary climatic and biological record from the seabed near the North Pole shows that 55 million years ago the Arctic Ocean was much warmer than scientists imagined — a Floridian year-round average of 74 degrees.

The findings, published today in three papers in the journal Nature, fill in a blank spot in scientists' understanding of climate history. And while they show that much remains to be learned about climate change, they suggest that scientists have greatly under-estimated the power of heat-trapping gases to warm the Arctic.

Previous computer simulations, done without the benefit of seabed sampling, did not suggest an ancient Arctic that was nearly so warm, the authors said. So the simulations must have missed elements that lead to greater warming.

"Something extra happens when you push the world into a warmer world, and we just don't understand what it is," said one lead author, Henk Brinkhuis, an expert on ancient Arctic ecology at the University of Utrecht in the Netherlands.

The studies draw on the work of a pioneering 2004 expedition that defied the Arctic Ocean ice and pulled the

Photo by John Farrell/University of Rhode Island
Researchers drilled cores from a largely unexamined seabed.

first significant samples from the ancient layered seabed 150 miles from the North Pole: 1,400 feet of slender shafts of muck, fossils of ancient organisms and rock representing a climate history that dates back 56 million years.

While there is ample fossil evidence around the edges of the Arctic Ocean showing great past swings in climate, until now the sediment samples from the undersea depths had gone back less than 400,000 years.

The new analysis confirms that the Arctic Ocean warmed remarkably 55 million years ago, which is when many scientists say the extraordinary planetwide warm-up called the Paleocene Eocene Thermal Maximum must have been caused by an enormous outburst of heat-trapping, or greenhouse, gases like methane and carbon dioxide. But no one has found a clear cause for the gas

discharge. Almost all climate experts agree that the present-day gas buildup is predominantly a result of emissions from smokestacks, tailpipes and burning forests.

The samples also chronicle the subsequent cooling, with many ups and downs, that the researchers say began about 45 million years ago and led to the cycles of ice ages and brief warm spells of the last several million years.

Experts not connected with the studies say they support the idea that heat-trapping gases — not slight variations in Earth's orbit — largely determine warming and cooling.

"The new research provides additional important evidence that greenhouse-gas changes controlled much of climate history, which strengthens the argument that greenhouse-gas changes are likely to control much of the climate future," said one such expert, Richard B. Alley, a geoscientist at Pennsylvania State University.

The $12.5 million Arctic Coring Expedition, run by a consortium called the International Ocean Drilling Program, was the first to drill deep into the layers of sediment deposited over millions of years in the Arctic. The samples were gathered late in the summer of 2004 as two icebreakers shattered huge drifting floes so that a third ship could hold its position and bore for core samples.

A Freshwater Arctic

After analyzing a sediment core from the Arctic Ocean seafloor, scientists theorize that the ocean was nearly closed off by the continents about 49 million years ago, creating a great pool of fresh water.

Source: Nature

The New York Times

Photo by Martin Jakobsson

The ship Vidar Viking had to remain stationary for days at a time amid drifting sea ice near the North Pole.

Estimates of the prevailing temperatures in the different eras represented by the sediments were made in part by tracking the comings and goings of certain algae called dinoflagellates that typically indicate subtropical or tropical conditions.

Because the samples lacked remains of shell-bearing plankton that are usually relied on to provide temperature records, the researchers used a newer method for approximating past temperatures: gauging changes in the chemical composition of the remains of a primitive phylum of microbes called Crenarchaeota.

Some scientists familiar with the research said that while there were still questions about the precision of this method at temperatures like those in the ancient Arctic Ocean, it was clear that the area was warm.

The temperatures recorded in the samples, right through the peak of warming 55 million years ago, were consistently about 18 degrees higher than those projected by computer models trying to "backcast" what the Arctic was like at the time, according to one of the papers.

Another significant discovery came in layers from 49 million years ago, where conditions suddenly fostered the summertime growth of vast mats of an ancient cousin of the Azolla duckweed that now cloaks suburban ponds. The researchers propose that this occurred when straits closed between the Arctic Ocean and the Pacific and Atlantic Oceans.

The flow of water from precipitation and rivers created a great pool of fresh water, but about 800,000 years after the blossoming of duckweed began, it ended with a sudden warming of a few additional degrees. The researchers suggest that this signaled when shifting land formations reconnected the Arctic with the Atlantic, allowing salty, warmer water to flow in, killing off the weed.

The researchers said the sediments held hints that Earth's long slide to colder conditions, and the recent cycle of ice ages and brief thaws, began quite soon after the hothouse conditions 50 million years ago. A centerpiece of their argument is a single pebble, about the size of a chickpea, found in a layer created 45 million years ago.

The stone could have been deposited on the raised undersea ridge only if it had been carried overhead in ice, said Kathryn Moran, a chief scientist on the drilling project, who teaches at the University of Rhode Island.

The stone was probably embedded in an iceberg or perhaps a plate of sea ice that tore free from a gravelly shore. It sank as the ice melted or broke apart, Dr. Moran proposed. Such "dropstones" have long been used to date when an oceanic region has been ice covered or ice free.

The amount of ice-carried debris in the sediment layers began to increase about 14 million years ago, the scientists said. That is also about when the great ice sheet that now weighs down eastern Antarctica originated, Dr. Moran noted. In general, the results from the Arctic drilling project suggest that the cooling and ice buildup at both poles happened in relative lockstep.

This simultaneity tends to support the idea that the cooling was caused by a drop in concentrations of carbon dioxide and other heat-trapping gases, which mix uniformly in the global atmosphere, said Dr. Moran and other members of the team.

Julie Brigham-Grette of the University of Massachusetts, an expert in past Arctic climates who was not connected with the new studies, cautioned against giving too much significance to the single sample, and particularly the single stone from 45 million years ago.

Dr. Brigham-Grette said it was vital to try to mesh the new core results with data gathered around Arctic coasts, where there is plenty of evidence for warm conditions in at least some places as recently as 2.4 million years ago.

Despite her doubts, she said, the project was a stunning achievement.

"It's all very, very exciting to me, because now we can start to rewrite the history of the Arctic," Dr. Brigham-Grette said. "It's like working a giant landscape puzzle of 500 pieces. For a while we only had 100 pieces. Now we have 100 more, and the picture is getting clearer."

IN REVIEW

1. How were the new data about the past Arctic climate collected?

2. How warm was the Arctic in the distant past? Briefly summarize the climate changes through time revealed by the new data.

3. Why is it surprising to find that the Arctic was once so warm? How does this support the idea that greenhouse gas concentration plays a far bigger role in regulating global temperatures than other factors, such as changes in Earth's rotation and orbit?

4. Do you think that the discovery of a surprisingly warm Arctic in the past should make us more concerned or less concerned about the potential consequences of global warming today? Defend your opinion.

Antarctica is covered by so much ice that, if it all melted, sea level would rise by more than 200 feet. Fortunately, even the most pessimistic predictions of models of global warming do not suggest that such large-scale melting could occur any time soon. Nevertheless, some substantial change seems to be afoot in Antarctica, perhaps suggesting that global warming might affect that continent and sea level more than most scientists had assumed.

Antarctica, Warming, Looks Ever More Vulnerable

By Larry Rohter
The New York Times, **January 25, 2005**

Correction Appended

From an airplane at 500 feet, all that is visible here is a vast white emptiness. Ahead, a chalky plain stretches as far as the eye can see, the monotony broken only by a few gentle rises and the wrinkles created when new sheets of ice form.

Under the surface of that ice, though, profound and potentially troubling changes are taking place, and at a quickened pace. With temperatures climbing in parts of Antarctica in recent years, melt water seems to be penetrating deeper and deeper into ice crevices, weakening immense and seemingly impregnable formations that have developed over thousands of years.

As a result, huge glaciers in this and other remote areas of Antarctica are thinning and ice shelves the size of American states are either disintegrating or retreating—all possible indications of global warming. Scientists from the British Antarctic Survey reported in December that in some parts of the Antarctic Peninsula hundreds of miles from here, large growths of grass are appearing in places that until recently were hidden under a frozen cloak.

"The evidence is piling up; everything fits," Dr. Robert Thomas, a glaciologist from NASA who is the lead author of a recent paper on accelerating sea-level rise, said as the Chilean Navy plane flew over the sea ice here on an unusually clear day late in November. "Around the Amundsen Sea, we have surveyed a half dozen glaciers. All are thinning, in

Photo by British Antarctic Survey via Associated Press

The Larsen B ice shelf collapsed over a 35-day period early in 2002, losing more than a quarter of its total mass.

some cases quite rapidly, and in each case, the ice shelf is also thinning."

The relationship between glaciers (essentially frozen rivers) and ice shelves (thick plates of ice protruding from the land and floating on the ocean) is complicated and not fully understood. But scientists like to compare the spot where the "tongue" of a glacier flows to sea in the form of an ice shelf to a cork in a bottle. When the ice shelf breaks up, this can allow the inland ice to accelerate its march to the sea.

"By themselves, the tongue of the glacier or the cork in the bottle do not represent that much," said Dr. Claudio Teitelboim, the director of the Center

for Scientific Studies, a private Chilean institution that is the partner of the National Aeronautics and Space Administration in surveying the ice fields of Antarctica and Patagonia. "But once the cork is dislodged, the contents of the bottle flow out, and that can generate tremendous instability."

Glaciologists also know that by itself, free-floating sea ice does not raise the level of the sea, just as an ice cube in a glass of water does not cause an overflow as it melts. But glaciers are different because they rest on land, and if that vast volume of ice slides into the sea at a high rate, this adds mass to the ocean, which in turn can raise the global sea level.

Glaciers in Retreat

Antarctic researchers are using space-based laser and radar imaging to map and monitor the movement of its ice sheets. Below, ICESat, a NASA satellite, shoots laser pulses to measure ice height. These large-scale studies are complemented by aircraft measurements in regions of greatest change.

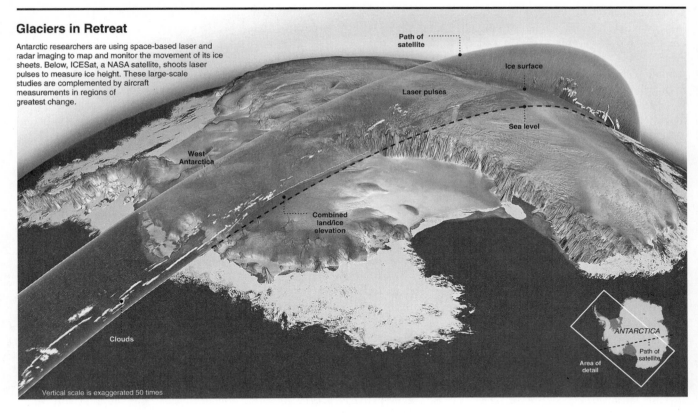

Western Ice Sheets Are Shrinking ...

Recent satellite data suggest that the Pine Island Glacier in West Antarctica is thinning rapidly. From 1992 to 2000, the grounding line - the line where ground-based ice turns to floating ice - retreated up to three miles.

    ⋯⋯ 1992   —— 2000

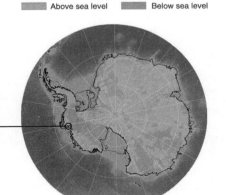

... Partly Because of Low Land Levels ...

While the East Antarctic ice sheet covers mostly land at or above sea level, much of the land under the western ice sheet is below sea level - making the ice susceptible to –accelerated flow into the ocean as its environment changes.

    ▨ Above sea level   ▨ Below sea level

... and Rising Temperatures

Although most of Antarctica has experienced cooling in the last 20 years, satellite data show warming along the coasts, which is believed to be weakening and thinning the ice.

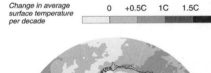

Change in average surface temperature per decade   0   +0.5C   1C   1.5C

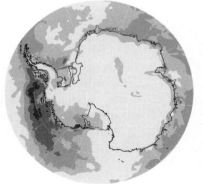

Sources: Waleed Abdalati, Dr, Josefino C. Comiso and the Scientific Visualization Studio, NASA Goddard Space Flight Center; Eric Rignot, NASA Jet Propulsion Laboratory; Canadian Space Agency

David Constantine/The New York Times

Through their flights over this and other areas of Antarctica, NASA and the Chilean center hope to help glaciologists and other scientists interested in climate change understand what is taking place on the continent and why. To do that, they need to compile data not only on ice thicknesses but also the underlying geology of the region, information that is most easily obtained from the air.

The flights are taking place aboard a Chilean Navy Orion P-3 plane that has been specially equipped with sophisticated instruments. The devices include a laser-imaging system that shoots 5,000 pulses of light per second at the ground to map the ice surface, as well as ice-penetrating radar to determine the depth of the ice sheets, a magnetometer and digital cameras.

For most parts of Antarctica, reliable records go back less than 50 years, and data from satellites and overflights like the ones going on here have been collected over only the past decade or so. But that research, plus striking changes that are visible to the naked eye, all point toward the disturbance of climate patterns thought to have been in place for thousands of years.

In 1995, for instance, the Larsen A ice shelf disintegrated, followed in 1998 by the collapse of the nearby Wilkins ice shelf. Over a 35-day period early in 2002, at the end of the Southern Hemisphere summer, the Larsen B ice shelf shattered, losing more than a quarter of its total mass and setting thousands of icebergs adrift in the Weddell Sea.

"The response time scale of ice dynamics is a lot shorter than we used to think it was," said Dr. Robert Bindschadler, a NASA scientist who is director of the West Antarctic Ice Sheet Initiative. "We don't know what the exact cause is, but what we observe going on today is likely to be what is also happening tomorrow."

Thus far, all of the ice shelves that have collapsed are located on the Antarctic peninsula. In reality a collection of islands, mountain ranges and glaciers, the peninsula juts northward toward Argentina and Chile and is "really getting hot, competing with the Yukon for the title of the fastest warming place on the globe," in the words of Dr. Eric Steig, a glaciologist who teaches at the University of Washington.

According to a recent study published in *Geophysical Research Letters*, the discharge rate of three important glaciers still remaining on the peninsula accelerated eightfold just from 2000 to 2003. "Ice is thinning at the rate of tens of meters per year" on the peninsula, with glacier elevations in some places having dropped by as much as 124 feet in six months, the study found.

But the narrow peninsula contains relatively little inland ice. Glaciologists are more concerned that they are now beginning to detect similar signs closer to the South Pole, on the main body of the continent, where ice shelves are much larger—and could contribute far more to sea level changes. Of particular interest is this remote and almost in-

accessible region known as "the weak underbelly of West Antarctica," where some individual ice shelves are as large as Texas or Spain and much of the land on which they rest lies under sea level.

"This is probably the most active part of Antarctica," said Dr. Eric Rignot, a glaciologist at the Jet Propulsion Laboratory in Pasadena, Calif., and the principal author of the Geophysical Research Letters paper. "Glaciers are changing rapidly and increasingly discharging into the ocean, which contributes to sea level rise in a more significant way than any other part of Antarctica."

According to another paper, published in the journal *Science* in September, "the catchment regions of Amundsen Sea glaciers contain enough ice to raise sea level by 1.3 meters," or about four feet. While the current sea level rise attributable to glacier thinning here is a relatively modest 0.2 millimeters a year, or about 10 percent of the total global increase, the paper noted that near the coast the process had accelerated and might continue to do so.

As a result, the most recent flights of NASA and the Chilean center have been directed over the Thurston Island and Pine Island zones of West Antarctica, near the point where the Bellingshausen and Amundsen Seas come together. The idea is to use the laser and radar readings being gathered to establish a base line for comparison with future measurements, to be taken every two years or so.

"We're not sure yet how to connect what we see on the peninsula with what we observe going on further south, but both are very clearly dramatic and dynamic events," Dr. Bindschadler said. "On the peninsula, large amounts of melt water are directly connected to disintegration of the ice shelf, but the actual mechanism in West Antarctica, whether melt water, a slippery hill or a firmer bedrock, is not yet clear. Hence the need for more data."

The information being gathered here coincides with the recent publication of a report on accelerating climate change in the Arctic, an area that has been far more scrutinized than Antarctica. That

study, commissioned by the United States and seven other nations, found permafrost there to be thawing and glaciers and sea ice to be retreating markedly, raising new concerns about global warming and its impact.

"The Arctic has lots of land at high latitudes, and the presence of land masses helps snow melt off more quickly," said Dr. Steig. "But there's not a lot of land to speak of in the high latitudes of the Southern Hemisphere," making the search for an explanation of what is going on here even more complicated.

The hypotheses scientists offer for the causes of glacier and ice shelf thinning in Antarctica are varied. Rising air, land and ocean temperatures or some combination of them have all been cited.

Some scientists have even proposed that a healing of the seasonal ozone hole over the South Pole and southernmost Chile, a phenomenon expected to take place in the next 50 years or so, could change the circulation of the atmosphere over the frozen continent in ways that could accelerate the thinning of Antarctic ice fields. But even without that prospect, the situation developing in Antarctica is already sobering, glaciologists agree. The data being collected here in West Antarctica and on the peninsula farther north make that obvious, they say, though the degree to which that should be cause for concern around the rest of the planet will become clear only with more research.

"If Antarctica collapses, it will have a major effect on the whole globe," Dr. Rignot cautioned. He warned that "this is not for tomorrow, and Antarctica is such a big place that it's important to look at other areas" around the perimeter of the giant continent, but added, "Nature is playing a little experiment with us, showing us what could happen if the plug were to be removed."

Correction: January 31, 2005, Monday. An article in *Science Times* on Tuesday about the effects of warming in Antarctica misidentified a feature of the landscape there. What is as large as Texas and sits on land that lies below sea level is ice sheets, not ice shelves.

IN REVIEW

1. List the changes that are happening in Antarctica due to increases in average surface temperatures on the continent.

2. Explain how warming in Antarctica might increase the average sea level on Earth.

3. Describe how scientists are measuring changes in the amount of water ice in Antarctica.

4. In what ways are the findings from Antarctica surprising to scientists studying global warming?

5. Does the information in this article alter your opinion in any way about what we should do, if anything, about the issue of global warming? Explain.

This commentary, written by a Times reporter who covers astronomy and physics, is about a recent trend in NASA to review and edit science stories for "controversial" topics. Although it is essentially an opinion piece, it touches on many topics from your textbook, including our understanding of the Sun, the evolution of stars like the Sun, and how we study stars and test our theories about them.

Someday the Sun Will Go Out and the World Will End (but Don't Tell Anyone)

By Dennis Overbye
The New York Times, **February 14, 2006**

I've always been proud of my irrelevance.

When I raised my hand to speak at our weekly meetings here in the science department, my colleagues could be sure they would hear something weird about time travel or adventures in the fifth dimension. Something to take them far from the daily grind. Enough to taunt the mind, but not enough to attract the attention of bloggers, editors, politicians and others who keep track of important world affairs.

So imagine my surprise to find the origin of the universe suddenly at the white hot center of national politics. Last week my colleague Andrew Revkin reported that a 24-year-old NASA political appointee with no scientific background, George C. Deutsch, had told a designer working on a NASA Web project that the Big Bang was "not proven fact; it is opinion," and thus the word "theory" should be used with every mention of Big Bang.

It was not NASA's place, he said in an e-mail message, to make a declaration about the origin of the universe "that discounts intelligent design by a creator."

In a different example of spinning science news last month, NASA headquarters removed a reference to the future death of the sun from a press release about the discovery of comet dust around a distant star known as a white dwarf. A white dwarf, a shrunken dense cinder about the size of earth, is how our own sun is fated to spend eternity, astronomers say, about five billion years from now, once it has burned its fuel.

"We are seeing the ghost of a star that was once a lot like our sun," said Marc Kuchner of the Goddard Space Flight Center. In a statement that was edited out of the final news release he went on to say, "I cringed when I saw the data because it probably reflects the grim but very distant future of our own planets and solar system."

An e-mail message from Erica Hupp at NASA headquarters to the authors of the original release at the Jet Propulsion Laboratory in Pasadena, Calif., said, "NASA is not in the habit of frightening the public with doom and gloom scenarios."

Never mind that the death of the sun has been a staple of astronomy textbooks for 50 years.

Dean Acosta, NASA's deputy assistant administrator for public affairs, said the editing of Dr. Kuchner's comments was part of the normal "give and take" involved in producing a press release. "There was not one political person involved at all," he said.

Personally, I can't get enough of gloom-and-doom scenarios. I'm enchanted by the recent discovery, buttressed by observations from NASA's Hubble Space Telescope, that an anti-gravitational force known as dark energy might suck all galaxies out of the observable universe in a few hundred billion years and even rip apart atoms and space. But I never dreamed that I might be frightening the adults.

What's next? Will future presidential candidates debate the ontological status of Schrödinger's cat? That's the cat that, according to the uncertainty principle of quantum physics, is both alive and dead until we observe it.

Apparently science does matter.

Dreading the prospect that they too may be dragged into the culture wars, astronomers have watched from the sidelines in recent years as creationists in Kansas and Pennsylvania challenged the teaching of evolution in classrooms. Never mind that the Big Bang has been officially accepted by the Roman Catholic Church for half a century. The notion of a 14-billion-year-old cosmos doesn't fit if you believe the Bible says the world is 6,000 years old.

And indeed there have been sporadic outbreaks, as evidenced by the bumper stickers and signs you see in some parts of the country: "Big Bang? You've got to be kidding — God."

When the Kansas school board removed evolution from the science curriculum back in 1999, they also removed the Big Bang.

In a way, the critics have a point. The Big Bang is indeed only a theory, albeit a theory that covers the history of creation as seamlessly as could be expected from the first fraction of a second of time until today. To call an idea "a theory" is to accord it high status in the world of science. To pass the bar, a theory must make testable predictions — that stars eventually blow out or that your computer will boot up.

Sometimes those predictions can be, well, a little disconcerting. When you're talking about the birth or death of the universe, a little denial goes a long way.

That science news is sometimes managed as carefully as political news may not come as a surprise to most adults. After all, the agencies that pay for most scientific research in this country have billion-dollar budgets that they have to justify to the White House and the Congress. It helps to have newspaper clippings attesting to your advancement of the president's vision.

It's enough to make you feel sorry for NASA, whose very charter man-dates high visibility for both its triumphs and its flops, but which has officers recently requiring headquarters approval before consenting to interviews with the likes of me.

The recent peek behind the curtains of this bureaucracy has been both depressing and exciting. So they are paying attention after all.

They should be paying attention, but I'm not looking forward to having to include more politicians and bureaucrats in my rounds of the ever-expanding, multi-dimensional universe (or universes).

I'll do it, but, lacking the gene for street smarts, I fear being played like a two-bit banjo. I'm even happy to go star-gazing with Dick Cheney, if duty so calls, but only if he agrees to disarm and I can wear a helmet.

IN REVIEW

1. Do you consider the prediction that the Sun will eventually "go out" to be controversial? Defend your opinion.

2. What is a theory, as a scientist uses the word? What is a theory, as a typical American uses the word? What is the difference between these two definitions? How does this difference sometimes lead to conflict between scientists and the general public?

3. The average astronomy course is full of gloom-and-doom scenarios, ranging from the fate of the Earth, to the fate of the Sun, to the fate of the Universe. Which ones (if any) are truly scary to you? Why?

For more than three decades, the "solar neutrino problem" was one of the great mysteries of the Sun. As discussed in the text, it now appears to have been solved, thanks to data collected by the Sudbury Neutrino Observatory (SNO) and other neutrino detectors. With this success in the bag, scientists are now looking to learn much more from neutrinos—not just about the Sun, but about processes throughout the universe.

Tiny, Plentiful and Really Hard to Catch

By Kenneth Chang
The New York Times, **April 26, 2005**

An hour north of Duluth, Minn., and a half-mile down, the dim tunnels of the Soudan mine open up to a bright, comfortably warm cavern roughly the size of a gymnasium, 45 feet high, 50 feet wide, 270 feet long.

Well hidden from the lakes, pine forests and small towns of northern Minnesota, the mine churned out almost pure iron ore until it closed in 1962. Today, it is a state park, and it houses a $55 million particle physics experiment that is part of a worldwide effort to unravel the secrets of the neutrino, one of the least known and most common elementary particles.

Because of discoveries over the past decade, the ubiquitous neutrino, once a curiosity in a corner of particle physics, now has the potential to disrupt much of what physicists think they know about the subatomic world. It may hold a key to understanding the creation of hydrogen, helium and other light elements minutes after the Big Bang and to how dying stars explode.

The experiment at Soudan will measure the rate that neutrinos seemingly magically change their types, giving physicists a better idea of the minute mass they carry. An experiment at Fermilab outside Chicago is looking for a particle called a "sterile neutrino" that never interacts with the rest of the universe except through gravity.

Astrophysicists are building neutrino observatories in Antarctica and the Mediterranean, which will provide new views of the cosmos, illuminating the violent happenings at the centers of galaxies, distant bright quasars and elsewhere.

The particle is nothing if not elusive. In 1987, astronomers counted 19

Photo by Sudbury Neutrino Observatory.

The Sudbury Neutrino Observatory in Ontario is much larger than its counterpart in northern Minnesota.

neutrinos from an explosion of a star in the nearby Large Magellanic Cloud, 19 out of the billion trillion trillion trillion trillion neutrinos that flew from the supernova. The observation confirmed the basic understanding that supernovas are set off by the gravitational collapse of stars, but there were not enough data to discern much about the neutrinos.

The much larger detectors in operation today, Super-Kamiokande in Japan, filled with 12.5 million gallons of water, and the Sudbury Neutrino Observatory in Canada, would capture thousands of neutrinos from a similar outburst.

Because neutrinos are so aloof, successful experiments must have either a lot of neutrinos, produced en masse by accelerators or nuclear reactors, or a lot of matter for neutrinos to run into.

Given the cost of building huge detectors, scientists are now turning to places where nature will cooperate.

In Antarctica, the IceCube project will consist of 80 strings holding 4,800 detectors in the ice, turning a cubic kilometer of ice into a neutrino telescope. Fourteen European laboratories are collaborating on a project called Antares that will similarly turn a section of the Mediterranean off the French Riviera into a neutrino detector.

The Soudan experiment takes the other approach, using bountiful bursts of neutrinos generated by a particle accelerator. Shoehorned into the back of the underground cavern is a detector of modest size, a mere 6,000 tons, consisting of 486 octagonal steel plates standing upright like a loaf of bread. Each plate,

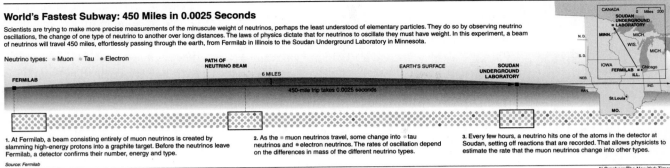

World's Fastest Subway: 450 Miles in 0.0025 Seconds

Scientists are trying to make more precise measurements of the minuscule weight of neutrinos, perhaps the least understood of elementary particles. They do so by observing neutrino oscillations, the change of one type of neutrino to another over long distances. The laws of physics dictate that for neutrinos to oscillate they must have weight. In this experiment, a beam of neutrinos will travel 450 miles, effortlessly passing through the earth, from Fermilab in Illinois to the Soudan Underground Laboratory in Minnesota.

1. At Fermilab, a beam consisting entirely of muon neutrinos is created by slamming high-energy protons into a graphite target. Before the neutrinos leave Fermilab, a detector confirms their number, energy and type.

2. As the ● muon neutrinos travel, some change into ● tau neutrinos and ● electron neutrinos. The rates of oscillation depend on the differences in mass of the different neutrino types.

3. Every few hours, a neutrino hits one of the atoms in the detector at Soudan, setting off reactions that are recorded. That allows physicists to estimate the rate that the muon neutrinos change into other types.

Source: Fermilab

Al Granberg/The New York Times

1 inch thick and 30 feet wide, weighs 12 tons.

On a visit to the cavern last month, William H. Miller, the laboratory manager, pointed at the far rock wall. "Fermilab, that way," he said. This experiment is intended to catch just a few of the neutrinos created at Fermilab, 450 miles away, which gush out of the rock wall, through the cavern, through the steel plates and then through another several miles of rock before emerging out of the earth and continuing into outer space, having no effect on Dr. Miller or the reporter interviewing him.

"You need a light-year of lead to reliably stop" a neutrino, said Dr. Alec T. Habig, a professor of astrophysics at the University of Minnesota at Duluth and the operations manager for the neutrino detector.

Only occasionally, a neutrino runs into a proton or neutron among the many atoms in the steel plates, and the wreckage of that collision is recorded as tiny bursts of light careering through the detector.

When the experiment begins running at full speed later this year, Fermilab will send trillions of neutrinos every couple of seconds flying toward Soudan. The beam will spread out to half a mile wide by the time it reaches Minnesota, so most of the neutrinos will miss the cavern entirely. But even among those that strike the bull's-eye, only one every few hours will actually hit something in the detector and be detected.

So far, in the testing phase in the past two months, the Soudan detector has seen just three, maybe four, neutrinos from Fermilab. But then, Wolfgang

Pauli thought physicists would never see any. Pauli, a pioneer of quantum theory, contrived the notion of neutrinos in 1930 to explain the disappearance of energy when unstable atoms fell apart. Pauli said the missing energy was being carried away by an unseen particle.

"I've done a terrible thing," Pauli wrote. "I have postulated a particle that cannot be detected." Pauli even wagered a case of Champagne that his particle would not be detected. In 1956, Pauli sent a case of Champagne to Clyde L. Cowan Jr. and Frederick Reines, two American physicists who proved him wrong using the flood of neutrinos produced in a nuclear reactor.

Physicists later discovered that neutrinos come in three types, whimsically called flavors. The flavor seen first was the electron neutrino, which interacts only with electrons. Heavier electron-like particles known as muons and tau particles are accompanied by their own flavors of neutrinos.

In 1998, an experiment at Super-Kamiokande showed that neutrinos change flavors as they travel along. For that to occur, the laws of physics dictate that the neutrinos, which had been thought be massless, must actually carry along a smidgen of weight, less than a millionth as much as an electron, the next lightest particle. Each flavor also has a slightly different mass.

In the Fermilab-to-Soudan experiment, the neutrinos are generated from a beam of protons, which are directed down a newly built $125 million tunnel, focused to a very narrow width with powerful electric fields and then slammed into a piece of graphite. That

produces short-lived particles called pions, which in turn generate muon neutrinos as they decay.

The beam passes through a smaller version of the Soudan detector, allowing the physicists to verify the number of neutrinos. The tunnel, sloped downward three degrees, ends just beyond the detector. The neutrinos keep going, into the earth, to emerge in the Soudan cavern one four-hundredth of a second later.

By using neutrinos created in an accelerator, physicists will be able to vary the energy of the neutrinos and see how that changes the number detected at Soudan.

"This wiggle has not really been seen," said Dr. Boris Kayser, a physicist at Fermilab. "It is one of the central expectations of our picture of neutrino oscillations."

The data from Soudan is expected to refine the Super-Kamiokande results, not overturn the prevailing wisdom. Another experiment at Fermilab may do just that.

A decade ago scientists at Los Alamos National Laboratory in New Mexico looked at neutrinos traveling a short distance from a nuclear reactor and saw indications of a large oscillation, suggesting a relatively large mass gap between two of the neutrino flavors. The gap was large enough that it could not fit into any theory consisting of just three neutrinos. But other experiments showed that only three flavors of neutrinos that interact with ordinary matter exist.

That has led to speculations of a new class of particles called sterile neutrinos. These particles would exert a force on other matter through gravity but would otherwise be completely inert.

"If it's really due to oscillations, then it implies physics way beyond the Standard Model," said Dr. William C. Louis of Los Alamos, who worked on the experiment.

Dr. Kayser said most theorists "are skeptical, because it doesn't fit," yet no one can point to obvious flaws in the Los Alamos work, either. The new experiment at Fermilab, called Mini-BooNE, is looking for the same effect but with a different setup, firing neutrinos into a spherical tank containing 250,000 gallons of baby oil.

(The name BooNE is an awkward contraction of Booster Neutrino Experiment. Booster refers to Fermilab's booster ring that accelerates protons, and the project's leaders added the prefix "mini" because they imagined a second, larger stage with a second detector if the current "mini" run confirms the Los Alamos findings.)

Dr. Louis, who is also one of the spokesmen for MiniBooNE, said that initial answers could be out by fall and insisted that he was not betting either way. "We're just concentrating on getting the correct result," he said, "and we'll worry about the consequences later."

The consequences may include the understanding of atom production in the aftermath of the Big Bang and in supernovas. Because neutrinos are essential to the nuclear reactions that change protons to neutrons and vice versa, they influence which elements form in what relative proportions.

"That would have profound implications for our models of the early universe and for supernovas," said Dr. George M. Fuller, a professor of physics at the University of California, San Diego. "It could change everything."

The problem is that current models that include three flavors of neutrinos do a good job of explaining the amount of hydrogen and helium in the universe. The existence of sterile neutrinos would send astrophysicists scurrying to come up with new calculations to produce the same answers.

On the other hand, current models of supernovas have trouble producing enough neutrons to form the heavier elements like uranium, and Dr. Fuller said sterile neutrinos could shift the reactions toward producing more neutrons.

Future experiments should aim at understanding other aspects of neutrinos, said Dr. Kayser, who was co-chairman of a committee that just released recommendations for future neutrino study.

For one, the neutrino oscillation findings say only that a difference in mass between the different flavors exists, but not the exact mass of any of them. The presumption is that because an electron is lighter than a muon and a muon is lighter than a tau that the same pattern should be true of the three neutrino flavors, with the electron neutrino the lightest and the tau neutrino the heaviest. But that does not have to be the case.

"It could be the other way around," Dr. Kayser said.

Physicists are also trying to learn whether an antineutrino is actually a neutrino. (Other antiparticles have opposite electrical charge. Because neutrinos are electrically neutral, nothing would prevent a neutrino from being its own antiparticle.)

Another open question is whether neutrinos play a role in the imbalance of matter and antimatter. If the early universe had contained equal amounts of the both, everything would have been annihilated, leaving nothing behind to form stars and galaxies.

Among quarks, which form protons and neutrons, physicists have observed a subtle matter-antimatter imbalance, called CP violation, in the behavior of particles known as mesons. "That CP violation is completely inadequate to explain the universe that we see," Dr. Kayser said.

So physicists suspect that there must be CP violation elsewhere and that the oddity of neutrinos suggest they could be a source. That, in turn, leads to speculation of yet more new types of neutrinos—very heavy ones that existed only in the very early universe—and the decay of those heavy neutrinos created the preponderance of matter.

Then come even wilder ideas—that neutrinos play a role in the mysterious dark energy that is pushing the universe apart or that neutrinos could be used for interstellar communication.

"Most of these ideas are of course probably wrong," said Dr. Louis of Los Alamos. "But if even one of them is right, it would be a tremendous breakthrough."

IN REVIEW

1. What are neutrinos? Why are they so difficult to catch? Why are neutrino detectors built deep underground?

2. Aside from solar emissions, what other cosmic events have produced neutrinos that we have detected in the past?

3. Briefly describe the Soudan neutrino experiment. What do scientists hope to learn from it? How will it complement other neutrino experiments, such as the Sudbury Neutrino Observatory?

4. How might a better understanding of neutrinos and observations of cosmic neutrinos lead to a better understanding of the early universe?

5. Until recently, we were able to study the distant cosmos only by observing light. How are neutrino observatories opening up a new window on the universe?

> As discussed in your text, observations set the mass limit of high-mass stars around 150 solar masses. This article reports on some of the observations that have led astronomers to conclude that this really is the mass limit for stars. The observations essentially rely on the study of a cluster with many stars, and the finding that none exceeded about 130 solar masses.

Stars on Diet: Weight Is Limited To 150 Suns, Researchers Find

By Warren E. Leary
The New York Times, **March 10, 2005**

WASHINGTON, March 9—The universe is full of stars, but there appear to be few really fat ones. Astronomers said Wednesday that there seemed to be a stellar weight limit equivalent to 150 Suns, but no bigger.

Using the Hubble Space Telescope to examine one of the densest clusters of stars in the Milky Way, which should have been brimming with fat stars, astronomers said they found a sharp cutoff in the mass of bodies that form in this stellar nursery.

In examining hundreds of stars in the dense Arches cluster, Dr. Donald F. Figer and colleagues at the Space Telescope Science Institute in Baltimore said they could not find any larger than 130 solar masses, or equal to the mass of 130 of our Suns.

"We are surprised at this result because we expected to find stars up to 500 to 1,000 times more massive than our Sun," Dr. Figer said.

At a telephone news conference organized by the National Aeronautics and Space Administration, experts called the findings a step to understanding star formation.

Dr. Sally Oey of the University of Michigan said that the findings, published in the March 10 issue of the journal Nature, were consistent with studies of smaller star clusters in our galaxy and observations she and colleagues had conducted of a huge star cluster in a galactic neighbor, the Large Magellanic Cloud.

The denser a cluster, the better the chance of finding giant stars, Dr. Oey said.

The Arches cluster that Dr. Figer examined, she said, "is the richest cluster in our galaxy." Because of this, astronomers said, it is highly unlikely that they would find superheavy stars elsewhere.

Astronomers have been uncertain about how massive a star can grow before it cannot hold itself together and blows apart.

Consequently, theories have predicted that stars can be 100 to 1,000 times more massive than the Sun. It has been easier to predict a lower weight limit for stars, experts said, because objects less than one-tenth the mass of our Sun are not heavy enough to sustain nuclear fusion in their cores to shine.

"These are fantastic findings," Dr. Stanford E. Woosley of the University of California, Santa Cruz, said of Dr. Figer's work.

Giant stars, at more than 100 solar masses, are important to galaxies and the universe because their furious combustion produces many important elements to form planets and other bodies like carbon, oxygen, sodium and neon, Dr. Woosley said.

The big stars also are short-lived, he said, with no star more than 100 solar masses lasting more than three million years because they consume their fuel so rapidly.

The Sun, by contrast, is 4.55 billion years old and expected to last 5 billion more years before running out of fuel. In mass, the Sun is equal to 300,000 planets the size of the Earth.

Dr. Figer said that although he found no star bigger than 130 solar masses in his observations, he set the upper limit for a big star at 150 solar masses to be conservative.

IN REVIEW

1. Suppose that Star cluster A has 1 million stars in it, while Star cluster B has 10,000 stars in it. Which cluster would you expect to have the most massive stars? Why?

2. Use your answer from Question 1 to explain why the observations reported in this article give astronomers confidence that they really have pinned down the mass limit for massive stars. (Hint: It may be helpful to consider this analogy: If you went to a small high school, you might find just a few 15-year-old boys with heights of 2 meters. At a larger high school, you might find some boys approaching 2.1 meters. But no matter what school you chose, you'd never find boys that are 3 meters tall, so you

could conclude that boys rarely if ever reach such heights.)

3. Why didn't astronomers already know the maximum mass of stars from theoretical considerations?

4. Why are very massive stars so rare? Why are they nonetheless important to our existence?

5. Do you think this result is the final scientific word on the topic of the existence of high-mass stars? Defend your answer.

The year 2006 marked the 1000th year anniversary of a supernova seen on Earth in 1006. Frank Winkler provides a timely essay reflecting on this event and our human connection to it.

Stardust Memories

By Frank Winkler
The New York Times, May 5, 2006

Middlebury, Vt.—HOW often do we get to celebrate the thousandth anniversary of anything? Younger readers may live to mark the millennium of the Battle of Hastings on Oct. 14, 2066, but surely none of us will be around on June 15, 2215, to celebrate the 1,000th anniversary of King John placing his seal on the document that became the Magna Carta.

This week, however, marks the millennium of a significant event for astronomers. On May 1, 1006, a new star suddenly appeared in the southern constellation Lupus, the wolf. Within a few days it brightened and became what was probably the brightest star ever witnessed in recorded human history — an event that astronomers today recognize as a supernova, the cataclysmic explosion that marks the death of a massive star.

The true cause of such celestial events was not clear in 1006, and their interpretation was the province of court astronomers, who served as astrologers as well. One of these was Zhou Keming, who wisely declared that the star's brilliance and golden color were portents of good fortune for the land where it appears — and received a promotion from the Chinese emperor.

Observers throughout the world recorded this dramatic event. In the Middle East and Africa, astronomers in Antioch, Alexandria, Cairo and Bagh-dad compared it with Venus, and even with the Moon, in brightness.

At the Abbey of St. Gall in Switzerland, where the star could barely be seen over the southern Alpine horizon, chroniclers nevertheless described it as the most significant event of the year: "a star of unusual magnitude, shimmering brightly in the extreme south, beyond all the constellations." And the Japanese poet Fujiwara Teika two centuries later celebrated the fame of the "great guest star" in his "Diary of the Clear Moon."

Aided by historical records, modern astronomers have identified what remains of the 1006 supernova today — a faint shell of gas about 7,000 light-years away. While we cannot be certain just how bright it appeared 1,000 years ago, opinion is virtually unanimous that no other star in recorded history was as bright.

Readers with reasonably dark skies can get an idea of how bright the star would have appeared in 1006 by a simple comparison. The brightest object (after the Moon) in the current evening sky is the planet Jupiter, low in the southeast. Just below it, by about the width of a finger held at arm's length, is a relatively faint star. At its brightest, the 1006 supernova would have been as much brighter than Jupiter as Jupiter is compared with that faint star.

It would have been bright enough to read by (for those few who could read in the 11th century) and could be seen even in daytime for weeks. Like all supernovae, it gradually faded, but remained visible for at least two and a half years, according to Chinese records.

Certainly these are rare events; the most recent in our own Milky Way galaxy to have been unquestionably visible to the naked eye was in 1604, just five years before Galileo first trained his telescope on the heavens. The only supernova since 1604 bright enough to be seen without a telescope was one visible to observers in the Southern Hemisphere in February 1987. It occurred not in the Milky Way, however, but about 170,000 light-years away in the Large Magellanic Cloud, our nearest neighbor galaxy.

Humans have particular reason to celebrate this anniversary. Within a supernova's fires of destruction are forged chemical elements that may eventually be incorporated into new stars, and planets and their inhabitants. Us, for instance. Most of the atoms in our bodies — the oxygen we breathe, the calcium in our bones, the iron in our hemoglobin — all stem from supernovae that occurred billions of years ago. And so, on the 1,000th anniversary of this stellar event, you might take a moment to get in touch with your cosmic roots and reflect that we are, in a very real sense, children of the stars.

IN REVIEW

1. Briefly summarize some of the known historical records of the supernova of 1006.

2. How bright was the supernova in the skies when it was seen in 1006?

3. Did the observers of 1006 know they were witnessing the explosion of a star? What did they think of the event?

4. Have any supernovae been bright enough to see with the naked eye in modern times? Explain.

5. Winkler concludes his article by stating that we are "children of the stars." What does he mean?

NASA's space telescopes not only produce great science, but they also occasionally produce iconic images that capture the imagination. Stars form in giant molecular clouds, and this process is anything but quiescent and dull. The drama of a star-forming region can be as gripping as a work of art, as new stars condense and light up in cocoons of molecular hydrogen, while their massive, short-lived siblings blast away their surroundings with radiation and stellar winds.

Note: The "Pillars of Creation" figure referred to in the article appears in your textbook as Figure 19.11 (in *The Cosmic Perspective*) or Figure 14.10 (in *The Essential Cosmic Perspective*).

Seeing Mountains in Starry Clouds of Creation

By Dennis Overbye
The New York Times, **November 15, 2005**

In 1995, astronomers using the Hubble Space Telescope produced "The Pillars of Creation," an image of stars emerging from biblical-looking clouds of dust that has become an icon of the space age.

Now astronomers operating NASA's Spitzer Space Telescope have made their own version. The new image, above, appropriately called "Mountains of Creation," shows star-forming pillars in a region known as W5 in the constellation Cassiopeia. These pillars, at heights up to 40 light-years, are 10 times as large as those in the famous Hubble image.

The astronomers, led by Lori E. Allen of the Harvard-Smithsonian Center for Astrophysics, say the towering mountains of the new image probably represent the densest, most fecund remnants of a larger, cloud. It is being eroded by radiation and winds of particles from a ferociously bright star just out of the top of the picture.

Nestled within the dusty pillars are hundreds of embryonic stars. But Spitzer's detectors are designed to see infrared, or "heat," radiation right

Photo by NASA

An infrared image of emerging stars captured by the Spitzer Space Telescope.

through the dust, allowing astronomers to study the cloaked stars, which Dr. Allen described as "offspring" of the big star.

"The Sun could have formed in such a cluster, since many stars form in clusters," Dr. Allen said in an e-mail message, explaining that pressure created by the star could compress gas in the cloud, bringing about the formation of new stars.

IN REVIEW

1. What kind of light does the Spitzer Space Telescope study? How does this light compare to the light that the Hubble Space Telescope sees? Compare its typical wavelength, speed, and how easily it travels through dust and molecular gas.

2. What are the "Mountains of Creation" made out of, and what is happening inside of them?

3. Do you think the name given to the "Mountains of Creation" picture is appropriate? Explain.

4. What does "fecund" mean in the context used in the article? How can an interstellar cloud be fecund?

5. In your opinion, why are pictures like the "Mountains of Creation" and "Pillars of Creation" so appealing to people in general?

This *New York Times* article discusses Type 1a supernovas, which the textbook describes as "white dwarf supernovas." As discussed in the text, a white dwarf supernova is thought to occur in a binary star system in which one star is a white dwarf and the other star transfers mass to it. Once the mass of the white dwarf exceeds the limit of 1.4 times solar mass (which the article refers to as the "Chandrasekhar mass"), it explodes as a supernova.

Life-or-Death Question: How Supernovas Happen

By Dennis Overbye
The New York Times, **November 9, 2004**

Once a second or so, somewhere in the universe, a star blows itself to smithereens, blossoming momentarily to a brilliance greater than a billion suns.

Nobody understands how these events, among the most violent in nature, actually happen. But, until recently, that didn't much matter unless you were a practitioner of the arcane and messy branch of science known as nuclear astrophysics.

Lately, however, supernovas have become signal events in the life of the cosmos, as told by modern science.

Using a particular species of supernova, Type 1a, as cosmic distance markers, astronomers have concluded that a mysterious "dark energy" is wrenching space apart, a discovery that has thrown physics and cosmology into an uproar.

As a result, the fate of the universe—or at least our knowledge of it—is at stake, and understanding supernovas has become essential.

Astronomers are busy on many fronts trying to figure out the details of these explosions—scanning the skies to harvest more of them in the act, peering at the remains of ancient supernovas to seek a clue to their demise, harnessing networks of supercomputers to calculate moment by moment reactions in the heart of hell.

This has resulted recently in a kind of two-steps-forward, one-step-back progress, encouraging astronomers that they are on the right track, generally, with their theories, but at the same time underscoring complexities and baffling puzzles when it comes to pinning down the details of what happens in the explosions.

Last month members of an international team of astronomers led by Dr. Pilar Ruiz-Lapuente of the University of Barcelona announced that they had found a star speeding away from the site of a supernova blast seen in 1572 by the astronomer Tycho Brahe. This supernova, which appeared as a "new star" in the constellation Cassiopeia, was one of the earliest studied by astronomers, and helped shatter the Aristotelian notion that the heavens above the Moon were immutable.

The newly discovered star, presumably the companion of the star that exploded, supports a long-held notion that such explosions happen in double star systems when one star accumulating matter from the other reaches a critical mass and goes off like a bomb.

Meanwhile, members of a group of astrophysicists using a network of powerful supercomputers to simulate supernova explosions say they have succeeded for the first time in showing how such a star could blow up.

Over the course of 300 hours of calculation at the University of Chicago's Center for Astrophysical Thermonuclear Flashes, otherwise known as the Flash center, they watched bubbles of thermonuclear fury rise from the depths of the star like a deadly jellyfish and then sweep around the surface and collide in an apocalyptic detonation that Dr. Donald Lamb, a Chicago astrophysicist, called "totally bizarre and novel."

If true, the Chicago results could help explain not only how stars explode, but why the explosions are almost but not exactly alike, allowing astronomers to better calibrate their measurements of dark energy.

Many supernova experts said, however, that such computer simulations were more of a good start than a final answer. Dr. J. Craig Wheeler of the University of Texas called the Flash center work "a courageous calculation," but added that many details needed to be filled in. "I don't think this is the end of the story," he said. The story of Type 1a supernovas, experts have long agreed, begins with a dense cinder known as a white dwarf, composed of carbon and oxygen, which is how moderate-size stars like the Sun, having exhausted their thermonuclear fuels of hydrogen and helium, end their lives.

If it happens to be part of a double star system, the white dwarf can accumulate matter from its companion until it approaches a limit, known as the Chandrasekhar mass—about 1.4 times the mass of the Sun.

At that point, so the story goes, the pressure and density in the previously dead star will be great enough to reignite the star and thermonuclear reactions will ripple upward, transmuting the carbon and oxygen into heavier and heavier elements, ripping the white dwarf apart while its companion goes flying off.

A three-billion-degree bubble of thermonuclear hell mushrooms upward through a star in the early milliseconds of a supernova explosion. Sweeping around the star's surface, the bubble could collide with itself, setting off a fatal detonation.

Until recently, however, there was little evidence of this. Two white dwarfs could collide, for example, and blow up. In that case there would be no survivor.

Tycho Brahe's supernova has now offered new evidence for the former model, of the white dwarf bomb.

That supernova is one of the few of Type 1a's that have occurred in our own galaxy, and so astronomers have long sought to find its companion. That star, astronomers reasoned, should be zinging along relative to its neighbors, as a result of having been released, like a stone from a slingshot, from its orbit around the suddenly deceased white dwarf.

The site of the supernova explosion is marked today by a small scruff of X-rays and radio waves in the sky.

Near the center of this patch the team found a sunlike star moving three times as fast as it neighbors.

The star has the right characteristics to have been the one donating material to the white dwarf that exploded, but the identification is not ironclad, a team member, Dr. Alex Filippenko of the University of California at Berkeley, said, explaining in an e-mail message that "it is 'possible' that the star just happened to be zooming through that region and is unrelated to the supernova."

One far out possibility, the astronomers say, is that more observations will reveal the ashes of the supernova polluting the outer layers of the star. But that is probably too much to wish for, said Dr. Stan Woosley of the University of California at Santa Cruz, pointing out that the explosion might have blown the outer layer of the star, ashes and all, off into space.

"This star sat next to, and for a while inside the most powerful thermonuclear explosion in the universe, 2.5 million, trillion, trillion megatons," Dr. Woosley said.

But the details of that explosion, which happens invisibly in a second or so, are still a mystery.

The light show seen by astronomers comes from radiation released by radioactive nickel, which decays to cobalt, and then to iron over the days and months after the cataclysm, releasing gamma rays that strike the ashes of the shattered star and make them glow briefly brighter than a galaxy.

Because all Type 1a supernovas start from the same point, astronomers have tried to use them as cosmic geodetic markers, standard candles whose distances can be inferred from how bright they appear.

But the supernovas are not standard enough. They vary in their luminosities by about 40 percent, which is similar enough to prove that the expansion of the universe is speeding up and that dark energy exists, astronomers say, but not good enough to pin down crucial details about the strength of this strange force and how it may be changing over cosmic time, and thus whether the universe will ultimately rip apart or come together in a "big crunch."

In order to reduce the uncertainties in their measurements astronomers need to know how or whether to correct their observations for differences in things like the age and chemical composition of the parent white dwarfs.

The problem is that there are two ways for the star to burn: like a flame, which is called deflagration, and as an explosion, a detonation, in which the burning propagates as a shock wave.

And neither type of burning, by itself, can easily explain what astronomers have seen in supernova explosions.

The slow burn, deflagration, results in more of a fizzle than an explosion, they say. It does not produce enough nickel to generate the light seen by astronomers and leaves much of the star unburned. Moreover, the parts that are burned are all jumbled up, while supernovas in the sky appear to be nicely layered, with the densest elements, like iron and nickel in the center, and light ones like silicon, sulfur and magnesium on the outside.

If the supernova consists simply of a detonation, on the other hand, the star would all turn to nickel, and that would result in too much light.

As a result, in the last 10 years many theorists have adopted a "Goldilocks" model of the explosion, in which the star burns in the flame mode for a while, slowly expanding, and then detonates when the density of the star has fallen to the value to make the right amount of nickel.

"The porridge has to be just the right temperature," said Dr. Wheeler, who described recent three-dimensional simulations by Dr. Vadim Gamezo and Dr. Elaine Oran, both of the Naval Research Laboratory in Washington, and Dr. Alexei R. Khokhlov of the University of Chicago, as "state of the art."

None of these "delayed detonation" models explain why or when the star would detonate.

The scientists had to put that into the calculation by hand. Finding a natural trigger for the detonation is the "silver chalice" of our profession, Dr.

Wheeler said, adding that automobile companies spend millions on the problem of ignition in car cylinders.

This is where the Flash center calculations come in. "It turns out that you need walls to have an explosion," explained Dr. Lamb. But a star has no walls. So how does it explode?

The Flash group, led by Dr. Tomasz Plewa of Chicago and the Nicolaus Copernicus Astronomical Center in Warsaw, was investigating what would happen if the white dwarf began burning in a flamelike manner not exactly at its center—an unlikely event in the case of a real star subject to turbulence—but a bit off-center. In addition to Dr. Lamb, the group included Dr. Alan C. Calder of Chicago.

The result was a bubble of flame rising from the depths and then sweeping around the star to become its own wall, crashing into itself at a temperature of three billion degrees and crushing densities, enough, the Chicago physicists say, to trigger detonation.

"We watched with eyes agog and jaws dropped as the thing unfolded," Dr. Lamb said.

But whether nature really works this way or not, Dr. Lamb and others agree,

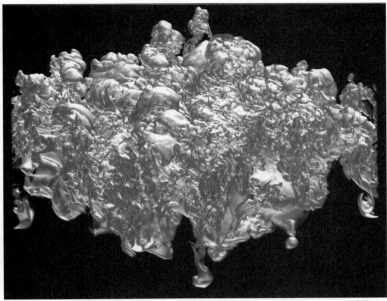

Photo by Center for Astrophysical Thermonuclear Flashes/University of Chicago

Gravity and buoyancy churn and warp the flame front in a star undergoing a supernova explosion. The front marks the boundary, as thick as a sheet of paper, where oxygen and carbon are being fused to heavier elements. Forty days and 40 nights on a supercomputer were required to produce this image representing a patch about half a yard across.

is yet to be determined, and it is far from a complete theory.

For one thing, the group has not yet been able to make three-dimensional calculations of the actual detonation. Such calculations could could be compared to observations.

As Dr. Woosley, said in an e-mail message, "just how a Type 1a supernova explodes is one of the most complicated things in the big wide world."

His group uses supercomputers to study small patches of the turbulent flame front—only a yard or two across— at high resolution.

Dr. David Arnett, a supernova expert at the University of Arizona, said

that such simulations were a way to test ideas and that watching them was a prod to theorists' intuition.

"Massive computing does not provide the answers so much as it provides an extension of our imagination," he wrote in an e-mail message. "For some years there has been talk of computing as being the third 'leg' of science: theory, experiment, computer simulations. I think the Flash work is a concrete example of this at work, and actually working."

Meanwhile, real supernovas threaten to confound the theorists. The carbon at the center of the star might "smolder" before it burns, putting it on a path to wind up as something other than nickel at the end, according to recent observations of two supernovas by Dr. Wheeler and his group using the Very Large Telescope at Le Serena, Chile. The evidence is sketchy, but that would mean that most of the models, including the Flash center's rising bubble, are wrong, he said.

But we shouldn't be discouraged. "We've come along way," Dr. Wheeler said. Referring to the ignition problem, he said, "We had to come a long way before we knew this was an issue."

IN REVIEW

1. A white dwarf supernova is one of two basic types of supernova; the textbook refers to the other type as a massive star supernova. How do the two types of supernova differ? Why do astronomers expect white dwarf supernovas all to be similar in luminosity, and why don't they expect the same for massive star supernovas?

2. How are white dwarf supernovas important to establishing the cosmic distance scale? List the main uncertainties associated with measuring the distances from Earth to these supernovas.

3. Describe the tools and observations scientists are using to unravel the mysteries associated with white dwarf supernovas.

4. How has the discovery of a star racing away from the site of Tycho's supernova given support to the idea that these supernovas represent the detonations of white dwarfs? Explain clearly.

5. How is a good understanding of these supernovas relevant to the question of the fate of the universe?

When supernova explosions occur, the resulting release of energy creates a flash at its peak luminosity as bright as 10 billion stars combined. But a supernova is not the "mother of all explosions." That title is reserved for gamma ray bursts, whose brightness may, for just an instant, exceed that of a million galaxies. The most recent of NASA's gamma-ray observatories, the Swift satellite, is designed to quickly detect a gamma ray burst and position itself to look squarely down its throat. Launched in November 2004, the Swift satellite has already made significant detections of gamma ray bursts, leaving high-energy astrophysicists salivating.

Dying Star Flares Up, Briefly Outshining Rest of Galaxy

By Kenneth Chang
The New York Times, February 20, 2005

For a fraction of a second in December, a dying remnant of an exploded star let out of a burst of light that outshone the Milky Way's other half-trillion stars combined, astronomers announced Friday.

Even on Earth, half a galaxy away, the starburst was one of the brightest objects ever observed in the sky, after the Sun and perhaps a few comets. The magnitude of the event caught most astronomers by surprise.

"Whoppingly bright," said Dr. Bryan M. Gaensler, an astronomer at the Harvard-Smithsonian Center for Astrophysics in Cambridge, Mass. "It gave off more energy in 0.2 seconds than the Sun does in 100,000 to 200,000 years."

No one on Earth directly saw the flare because most of the light was gamma rays, photons that are more energetic than X-rays and are blocked by the atmosphere. But the Dec. 27 pulse registered on instruments aboard 15 spacecraft, including NASA's new Swift satellite, which was designed to record cosmic gamma rays and had been turned on just the week before.

Dr. Neil Gehrels, the lead scientist for the Swift satellite, said flares of that magnitude could be expected just once in a millennium.

"That seems so improbable it's a puzzle right now," Dr. Gehrels said. "There's something going on here that we don't understand."

The radiation even temporarily compressed Earth's ionosphere, an envelope of charged gas at the top of the atmosphere, and distorted long-wavelength radio signals.

"It was really the big one," said Dr. Kevin Hurley, a researcher at the Space Sciences Laboratory at the University of California, Berkeley. "You could not have missed it."

Dr. Hurley and others pinpointed the origin of the pulse as a neutron star known as SGR 1806-20, about 50,000 light-years distant in the constellation Sagittarius. Neutron stars are remnants of stars after they have exploded in supernovas, and SGR 1806-20 is one of about 10 unusual neutron stars known as magnetars, which have extraordinarily strong magnetic fields, a quadrillion times as strong as Earth's.

Sudden shifts in the intense magnetic fields are believed to generate flares, in much the way the Sun generates solar flares, and two giant magnetar flares had been previously observed, one in 1979 and one in 1998. The Dec. 27 flare, however, was 100 times as powerful.

One physicist, Dr. David Eichler of Ben-Gurion University in Israel, wrote a paper in 2002 suggesting that magnetars might be capable of such cataclysmic flares, but most scientists, Dr. Gaensler said, "had no idea they could make a flare this big."

SGR 1806-20 has almost as much mass as 1.5 Suns, compressed into an incredibly dense ball about a dozen miles wide, and it spins around once every 7.5 seconds, slow for a neutron star. Discovered in 1979, SGR 1806-20 has at times been noisy, firing off small gamma ray flares, and at other times quiet. Its activity picked up in the past year.

"In retrospect, I guess you could say it was getting ready to let go," Dr. Hurley said, adding that he thought the magnetic fields, held in place by the crust of the star, had become twisted, building stress. "At some point, it gives way like an earthquake," he said.

Astronomers presented their observations at a NASA news conference on Friday, and several scientific papers describing the event will be published in a coming issue of the journal Nature.

In the aftermath of the flare, Dr. Gaensler, lead author of one of the Nature papers, and his colleagues used radio telescopes on Earth to track a shock wave radiating from the star. They were surprised to find the wave expanding rapidly, at a quarter of the speed of light, "which is not what you tend to see in the galaxy every day," he said.

SGR 1806-20 itself continues to spin as before, one revolution every 7.5 seconds. "Amazingly, the neutron star is still there," Dr. Gaensler said. "It did not explode or blow itself apart to bits."

The magnetar flare may help solve a cosmic mystery of gamma ray bursts, the prime mission of the Swift satellite. Bursts lasting from several seconds to a couple of minutes are believed to be caused by the collisions of black holes—events that are even more violent than magnetar flares and occur much farther away—but astronomers had been at a loss to explain shorter bursts lasting a couple of seconds or less.

Now they have at least a partial answer: some of the bursts are magnetar flares originating in other galaxies.

"It is clear magnetar flares make short gamma ray bursts," said Dr. Robert C. Duncan of the University of Texas. "It is at least a significant fraction of them."

But that still may not be the whole answer. Dr. Gehrels, the lead scientist for the Swift satellite, said that when astronomers looked in the direction of three recent short gamma ray bursts, those parts of the sky turned out to be empty.

"It just all fits so well, and then there were no galaxies there," Dr. Gehrels said.

But he said that as the satellite observed more gamma ray bursts, "we should know in the next couple of months the answer to this."

IN REVIEW

1. Discuss the event that happened in December 2004 that is described in the article. What mysteries are associated with this event? How has technology made the discovery of this event possible?

2. Describe the characteristics of magnetars. How do they differ from "regular" neutron stars?

3. What might have happened on magnetar SGR 1806-20 that resulted in an intense gamma ray burst? What surprising things were observed with this burst?

4. What uncertainties remain about the nature and origin of gamma ray bursts?

5. In your opinion, is it worthwhile to have orbiting observatories that can observe light wavelengths blocked by the Earth's atmosphere? In what ways, if any, have these orbiting observatories benefited humanity? In what ways, if any, have they been a burden on humanity? Discuss the benefits vs. the cost of space-based astronomy.

As discussed in your text, observational evidence points strongly to the existence of a 4-million-solar-mass black hole at the center of our Milky Way Galaxy. To date, the evidence is all indirect — we observe stars and other matter orbiting around the black hole, but our telescopes are not powerful enough to see the black hole itself. This article reports on progress toward reaching the point where we could see the black hole's "shadow" directly.

Astronomers Edging Closer To Gaining Black Hole Image

By Dennis Overbye
The New York Times, November 3, 2005

Astronomers are reporting today that they have moved a notch closer to seeing the unseeable.

Using a worldwide array of radio telescopes to obtain the most detailed look yet at the center of the Milky Way, they said they had determined that the diameter of a mysterious fountain of energy there was less than half that of Earth's orbit about the Sun.

The result strengthens the case that the energy is generated by a black hole that is gobbling stars and gas, they said. It also leaves astronomers on the verge of seeing the black hole itself as a small dark shadow ringed with light, in the blaze of radiation that marks the galaxy's center.

Until now, the existence of black holes — objects so dense that not even light can escape them—has been surmised by indirect measurements, say of stars or gas swirling in their grip. Seeing the black hole's shadow would require the ability to see about twice as much detail as can now be discerned. Such an observation could provide an important test of Albert Einstein's theory of general relativity, which predicts that black holes can exist.

"We're getting tantalizingly close to being able to see an unmistakable

Photo by Eric Agol/University of Washington
A simulated black hole, seen as a shadow with a halo of radiation.

signature that would provide the first concrete proof of a supermassive black hole at a galaxy's center," Shen Zhiqiang of the Shanghai Astronomical Observatory, a leader of an international team of radio astronomers, said in a news release. Their report appears today in the journal Nature.

Another member of the team, Fred K. Y. Lo, director of the National Radio Astronomy Observatory in Charlottesville, Va., said that achieving the extra resolution could take several years and would probably require new radio telescopes.

"We're not there yet," he said, "but in time, no question, we will get there."

He added that seeing the shadow would be "proof of the pudding" that Einstein was right.

In an accompanying commentary, Christopher Reynolds of the University of Maryland wrote that such observations would "herald a new era in probing the structure and properties of some of the most enigmatic objects in the universe."

But other experts said it might be difficult, even if the extra resolution could be achieved, to untangle the detailed properties of the black hole from its blazing surroundings.

Astronomers have identified thousands of probable black holes. The candidates include objects billions of times as massive as the Sun at the centers of galaxies, where, it is theorized, gas and dust swirling toward their doom are heated and erupt with jets of X-rays and radio energy.

But the putative holes are too far away for astronomers to discern what would be their signature feature: a point of no return called the event horizon, in effect an edge of the observable universe, from which nothing can return. Instead, the evidence for black holes rests mainly on the inference that too much invisible mass resides in too small a space to be anything else.

The center of the Milky Way is about 26,000 light-years away, in the direction of Sagittarius. The new observations conclude that at the center of the galaxy an amount of invisible matter equal to the mass of four million Suns is crammed into a region no more than 90 million miles across. That small size, the radio astronomers said, eliminates the most likely alternative explanation of the fireworks at the galaxy's center: a cluster of stars. Such a dense cluster would collapse in 100 years.

Even more conclusive proof would come from the observation of the black hole's shadow, which would be about five times the size of the event horizon and appear about as big as a tennis ball on the Moon as seen from Earth, according to calculations by Eric Agol of the University of Washington, Heino Falcke of the Max Planck Institute for Radio Astronomy in Germany and Fulvio Melia of the University of Arizona.

"For most people, seeing is believing," said Dr. Agol, who added that observations of the shadow could in principle be used to test whether general relativity is correct in such strange conditions and to measure how fast the black hole is spinning.

Martin Rees of Cambridge University in England, who with Donald Lynden-Bell in 1971 first proposed a black hole as the energy source at the Milky Way's center, said he was encouraged by this progress. But he cited studies suggesting that the shadow could be washed out by radiation or particles in front of the black hole, making definitive measurements hard.

As all the astronomers pointed out, getting to the next level of detail will require building new radio telescopes that operate at shorter wavelengths — and higher frequencies — than the Very Long Baseline Array of radio telescopes that were used to carry out the present observations.

"It's something I've been working on for 30 years," said Dr. Lo of the National Radio Astronomy Observatory. "It's been a long saga."

For a long time, he said, astronomers were peering through a haze. "Now we're seeing the thing in itself."

IN REVIEW

1. According to the new observations, what is the maximum size, in miles, of the object at the center of the Milky Way Galaxy? How does this size compare to the size of the Earth's orbit around the Sun?

2. Why do astronomers consider this new maximum size value, combined with the estimated mass of 4 million Suns, to provide very strong evidence that the central object is a black hole?

3. How were the new observations made, and how might they be further improved in the future?

4. What does the article mean when it says that we are getting closer to seeing the black hole's "shadow"? What kind of angular resolution would be necessary to see this shadow? Use an analogy to put this angular resolution in perspective.

5. Do you think direct observations of a black hole's shadow would make more people believe that black holes really exist? Defend your opinion.

Our Milky Way galaxy holds stars of many ages, including some that are as old as almost any stars in the universe. It also hosts what seems to be a very massive black hole in its center. This article surveys recent discoveries about the nature of the galaxy that may shed light on the evolution of the universe as a whole.

In Galaxies Near and Far, New Views of Universe Emerge

By John Noble Wilford
The New York Times, January 14, 2003

Some astronomers came and exulted over glimpses of objects so far away and thus so long ago, about 13 billion light-years, that they opened eyes and minds to the universe as it was soon after stars and galaxies first began popping up everywhere.

They focused on three quasars, luminous objects thought to be powered by massive black holes, from that early time; one was the most distant quasar ever observed. They stretched their horizons at the sight of a new image from the Hubble Space Telescope showing a glittering expanse of galaxies so remote that some of them had not been seen before; a few may be the earliest in the universe.

For other scientists here at the winter conference of the American Astronomical Society, there was astonishment and mystery enough closer to home, in Earth's own Milky Way galaxy. X-ray emissions revealed the messy eating habits of the black hole at the galactic center, and radio waves showed the magnetic field there to be surprisingly tangled.

The newest and most impressive discovery was made on the fringes of the Milky Way. Two teams of astronomers have observed a previously unseen band of hundreds of millions

Photo by Hubble Space Telescope/NASA

Acting as a cosmic lens in space, the gravity of a cluster of galaxies has amplified the light from galaxies near the dawn of time and smeared them into arcs in this picture taken by the Advanced Camera for Surveys on the Hubble Space Telescope and released last week. ABELL 1689, whose galaxies appear large and yellowish, is about 2.2 billion light-years away. DISTANT GALAXIES would be invisible without the amplification of the gravitational lens. ARCS AND DOTS One galaxy, about 12 billion light-years away, has been split by the lens into two images. Many arcs, representing galaxies at a range of about 10 billion light-years, form a partial bull's-eye pattern around the center of the cluster. By analyzing these images, astronomers can map the mass, including unseen dark matter in the cluster.

of stars orbiting beyond the galaxy's main disk, more than twice as far from the galactic center as the solar system.

The most likely explanation, astronomers say, is that the ring stars are remnants of smaller galaxies that came too close and were captured by the Milky Way's overpowering gravity.

This is more evidence, then, that in part the galaxy grew to its present size at the expense of less fortunate neighbors.

"This is a vivid smoking gun of the disruption of a satellite galaxy," said Dr. Bruce Margon of the Space Telescope Science Institute in Baltimore, who was not on either discovery team.

Part of the ring was observed by scientists from the Sloan Digital Sky Survey, a project for mapping one-quarter of the sky in three dimensions that is based at Sunspot, N.M. A team of Australian, British and Dutch astronomers, working at a telescope in the Canary Islands, then saw other sections, leading scientists to recognize that the band of stars appeared to reach around the entire galaxy.

The ring's encircling diameter is estimated at 120,000 light-years, the teams reported. Its thickness is about 10 times that of the rest of the galaxy, extending well above and below the galactic plane. Gravity, primarily from unseen dark matter of an unknown nature, holds the ring of up to 500 million stars—about the stellar population of small galaxies—in a nearly circular orbit.

Such a vast congregation of stars in a coherent ring had remained hidden from view because it lies in the same plane as the Milky Way disk and so was obscured by intervening stars, gas and dust. It was hard to distinguish the ring stars from the other matter and impossible to recognize their number or their organization in a discrete torus.

"Our entire picture of the Milky Way is being changed with this discovery," said Dr. Brian Yanny of the Fermi National Accelerator Laboratory outside Chicago, who was co-leader of the Sloan Survey group that detected the ring. Stars are more numerous and more closely spaced toward the center of galaxies and are expected to thin out more or less evenly in the outskirts.

The other leader, Dr. Heidi Jo Newberg of Rensselaer Polytechnic Institute in Troy, N.Y., said, "When we find large groups of stars formed into rings, it's an indication that at least part of our galaxy was formed by a lot of smaller or dwarf galaxies mixing together."

The Milky Way originated about 10 billion years ago, and it probably captured some neighbors several times since then, growing into a collection of about 400 billion visible stars. Other researchers at the meeting reported seeing a faint trail of stars that appear to be leftovers from a similar capture by Andromeda, a nearby galaxy not unlike the Milky Way.

But Dr. Annette Ferguson of the University of Groningen in the Netherlands, a member of the Canary Islands observing team, offered a possible alternative explanation for the ring. The stars there might have come from inside the galaxy's disc, she suggested, and gravitational interactions could have disturbed their orbits, causing them to migrate outward and into the ring.

In any case, scientists said, the ring promises to be an ideal place to study the mysterious dark matter, which makes up most of the mass of the universe, and its role in shaping cosmic structures.

Other researchers have been taking a closer, sharper look into the heart of the Milky Way, a region hidden from ordinary view by a fog of dust and gas. What they are finding in radio and X-ray observations surpasses immediate understanding.

The magnetic field near the galactic center, it seems, is more chaotic than previously thought.

Using the radio telescopes of the National Science Foundation's Very Large Array in New Mexico, a team of astronomers studied strange filaments produced by the interaction between the galaxy's magnetic field and high-velocity electrons. The filaments seen before were all aligned in nearly the same direction, like iron filings near a bar magnet.

But in the new observations, astronomers were surprised to see many more filaments oriented in various directions, tangled, they said, "like a bowl of spaghetti."

Dr. Joseph Lazio of the Naval Research Laboratory, a leader of the observing team, said, "The magnetic field in the center may not be all that organized or all that strong."

Another team leader, Dr. Namir Kassim, also of the Naval laboratory, could only conclude, "The Milky Way's center is an exciting, mysterious region that, once again, has given us a surprise."

And then there was the latest report on the eating habits of the black hole believed to lurk at the Milky Way's core.

A black hole, by definition, cannot be observed directly. It is an extremely compact massive object so dense that nothing, not even light, can escape its gravity. Scientists infer its presence by observing the turbulence and high-energy emissions its gravity creates on surrounding material.

Some black holes in other galaxies may weigh as much as several billion times the mass of the Sun, scientists think, but the Milky Way's is thought to be a puny three million times the solar mass.

In the longest and most sensitive X-ray observations made so far of the Milky Way's center, NASA's Earth-orbiting Chandra X-ray Observatory zoomed in on the region of the suspected black hole.

"We are getting a look at the everyday life of a black hole like never before," said Dr. Frederick K. Baganoff of the Massachusetts Institute of Technology. "We see it flaring on an almost daily basis."

What did this mean? The hot gases of these X-ray flares could be crumbs from the black hole's last meal of stellar matter, scientists said. Its table manners are atrocious.

The frequency of the X-ray flares suggests that the black hole eats often, scientists further said, but their weak intensity suggests that its meals are more like snacks than banquets.

"Although it appears to snack often, this black hole is definitely on a severe diet," Dr. Baganoff said. "This could be because explosive events in the past blew away much of the gas from the neighborhood of the black hole."

IN REVIEW

1. Describe each of the recent discoveries made about the Milky Way that are discussed in the article.

2. Explain how scientists discovered the ring of stars orbiting at the fringe of the Milky Way. Why had the ring been previously hidden?

3. How does the presence and structure of the ring give clues to how the Milky Way was formed? What uncertainties remain about the origin of the ring?

4. What are X-ray flares telling us about the black hole suspected to lie at the galactic center?

5. How do these new discoveries about our own galaxy bear on our understanding of the overall evolution of the universe?

In addition to the most famous space observatories such as the Hubble Space Telescope and Chandra X-ray Observatory, NASA and the space agencies of other nations have launched many smaller orbiting observatories. These smaller satellites have provided an incredible amount of data that have greatly increased scientific understanding of stars and galaxies. One such small satellite, the Galaxy Evolution Explorer (Galex), was launched in April 2003. By detecting ultraviolet light, the spacecraft's mission is to observe hundreds of thousands of galaxies to learn more about the characteristics of galaxies and, specifically, the younger stars inside them.

Three Dozen New Galaxies Are Found in Nearby Space

By Dennis Overbye
The New York Times, December 22, 2004

Fourteen billion years after the Big Bang started it all, there is still life in the old cosmos.

Astronomers announced yesterday that they had discovered three dozen baby galaxies in what passes for nearby space in the universe—two billion to four billion light-years distant. The galaxies, which are blossoming with new stars at a prodigious rate, resemble the infant Milky Way 10 billion years ago, the astronomers said.

Studying these new galaxies could give cosmologists new insights into the processes by which galaxies and stars first formed out of clouds of primordial gas and dust at the beginning of time.

"It's like looking out your window and seeing a dinosaur walk by," said Dr. Tim Heckman of Johns Hopkins University, who led a team using a NASA satellite, the Galaxy Evolution Explorer, or Galex, to pinpoint the newborns. Dr. Heckman spoke in Pasadena, Calif., at a news conference at the Jet Propulsion Laboratory, which manages the satellite. A paper describing the results has been submitted to The Astrophysical Journal.

The babies were a pleasant surprise.

Like the parents of a woman of a certain age who long ago gave up hope of grandchildren, astronomers had given up hope that the universe was still producing galaxies that could grow up to be the size of the Milky Way. The heyday of making stars, the active ingredients of galaxies, was five billion to

eight billion years ago. Perhaps only dwarf galaxies were being born today.

"We didn't know if there were any newborns still around or if this phase of cosmic creation is over," Dr. Heckman explained.

The baby galaxies appear as bluish blobs of light about 10,000 light-years across in images sent back by the Galex satellite, which was launched in 2003 on a 29-month mission to survey the sky for ultraviolet emissions.

Ultraviolet light, which has a shorter wavelength than visible light, is pro-

duced by the hottest, most massive stars, like those of the Pleiades cluster, which shines in the sky above Orion these frigid crystalline nights. Because such stars do not last very long, they are also among the youngest stars in the sky.

As a result, young galaxies stand out in ultraviolet light, said Dr. Chris Martin of the Jet Propulsion Laboratory, the principal investigator for the Galex project. "Ultraviolet traces star formation," Dr. Martin said.

The hitch for astronomers hoping to study the recent evolution of stars and

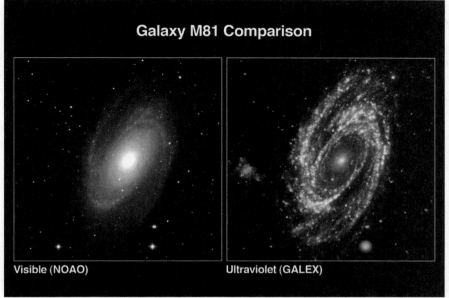

Galaxy M81 Comparison

Visible (NOAO)   Ultraviolet (GALEX)

Photo by National Optical Astronomy Observatory/NASA

The nearby galaxy Messier 81 as it appears in visible light and in ultraviolet light, which the NASA satellite Galaxy Evolution Explorer is using to map formation of galaxies across billions of years of cosmic time.

galaxies is that the atmosphere blocks ultraviolet rays from reaching Earth. So ultraviolet astronomy can be pursued only in space, with instruments like the Hubble Space Telescope and Galex.

Galex is designed to spot the ultraviolet glows of young stars and galaxies and thus help fill in the history of star formation and cosmic evolution over the last 10 billion years. It has a specially designed 20-inch-diameter telescope with a field of view four times as big as a full moon.

The new babies are only the first results of the project, and the astronomers said they expected to find more, although not many.

While they are not nearly the size of mature galaxies like the Milky Way, which is about 100,000 light-years across and has about 200 billion stars, the newborn galaxies outshine them in ultraviolet by a factor of 100 or so, which means they are producing stars "at a prodigious rate," in the words of Dr. Martin.

Dr. Alice Shapley, a theorist at the University of California, described them as "stragglers" of the great wave of galaxy formation that peaked when the universe was half its present age.

It is important, Dr. Shapley said, to try to find out what is finally causing these galaxies to form now. Are they accreting fresh star material from outside, for example? Indeed, she said, astronomers still do not know for sure whether these are really new galaxies, or whether perhaps they are old galaxies, hiding old stars inside them, that are undergoing a new burst of star formation.

These would be ideal objects to study with the Hubble Space Telescope, she added.

What will happen to these newborns is another mystery, Dr. Heckman said.

The infant Milky Way coalesced out of the murk 10 billion years ago, when the universe was more crowded and baby galaxies could bang into one another, merge and grow. "It's less clear what will happen in the future," Dr. Heckman said.

The universe is now a more diffuse place, and the baby galaxies may have been born into loneliness. If so, they will never grow up.

IN REVIEW

1. What are the characteristics of the infant galaxies discovered by Galex?

2. How far away are the newly discovered galaxies? How does their distance compare with the size of the observable universe? What does this suggest about how recently the galaxies were born? Can we be certain that they really are young galaxies?

3. In what way is the existence of these young galaxies surprising?

4. Why were these young galaxies easy to spot in ultraviolet light detected by Galex? Why is a satellite needed to detect UV light from the cosmos?

5. Why are these young galaxies rare in the universe? Explain why these small galaxies are not expected to grow to the size of our own galaxy, the Milky Way.

It has been more than 70 years since Edwin Hubble discovered the expansion of the universe. For most of the time since then, scientists assumed that the expansion should be slowing with time, as the mutual gravity of all the matter in the universe tries to pull itself back together. However, in the late 1990s, the observational data began to reveal a different scenario: Rather than slowing down, the expansion seems to be accelerating. New observations are gradually teaching us more about this apparent acceleration of the cosmos.

From Distant Galaxies, News of a "Stop-and-Go Universe"

By John Noble Wilford
The New York Times, **June 3, 2003**

New observations of exploding stars far deeper in space, astronomers say, have produced strong evidence that the proportions of the mysterious forces dominating the universe have undergone radical change over cosmic history.

The findings, reported here at a meeting of the American Astronomical Society, which ended Thursday, supported the idea that once the universe was expanding at a decelerating rate but then began accelerating within the last seven billion years, scientists concluded.

"We are now seeing hints that way back then the universe was slowing down," said Dr. John Tonry, an astronomer at the University of Hawaii who is a member of one team studying exploding stars, or supernovas, for signs of cosmic expansion rates.

The new research by Dr. Tonry's group and another, led by Dr. Saul Perlmutter of Lawrence Berkeley National Laboratory in California, confirmed the earlier surprising discovery that the universe is indeed expanding at an accelerating rate and has been for at least the last 1.2 billion years. But four supernovas, almost 7 billion light-years away, appeared to exist at a time the universe was slowing down, Dr. Tonry said.

"A stop-and-go universe" is the way Dr. Robert P. Kirshner of the Harvard-Smithsonian Center for Astrophysics characterized the phenomenon. Well, the expansion never really stopped,

NASA
A young supernova was recently detected in Arp 299, a colliding galaxy pair 140 million light-years away.

he conceded, but it has certainly revved up.

"Right now, the universe is speeding up, with galaxies zooming away from each other like Indy 500 racers hitting the gas when the green flag drops," said Dr. Kirshner, a member of the Tonry team. "But we suspect that it wasn't always this way."

The changing pace of cosmic expansion, combined with recently announced measurements of the cosmic microwave background, revealing conditions soon after the Big Bang, encourages theorists in thinking that a tug-of-war has been going on between dark forces of matter and energy no one yet understands.

The combined gravitational pull from all matter in the universe, most of which is beyond detection, has acted as a brake on cosmic expansion. The socalled dark matter apparently had the advantage when the universe was younger, smaller and denser. Now the ever-increasing pace of expansion suggests that something else even more mysterious is at work. Theorists are not sure what the antigravity force is, but they call it dark energy. It has apparently gained the upper hand.

This is the latest turn of events in the unfolding story of cosmic history. Once scientists believed the universe was everlastingly static. Along came Edwin P. Hubble, who discovered seven decades ago that the galaxies of stars are rushing away from one another in all directions. The universe, Hubble announced, is expanding.

Five years ago, astronomers were in for a surprise. They had assumed that after an initial burst of rapid expansion from the originating Big Bang the gravity of matter was gradually slowing things down. Then the two supernova

Changes of Pace

Scientists now believe that after about 7 billion years of slowing down, dark energy overtook dark matter and the universe began to accelerate.

| AGE OF UNIVERSE: | 1 MILLION YEARS | 500 MILLION | 1 BILLION | 7 BILLION | EARTH, SUN BORN: 9.35 BILLION | 14 BILLION (TODAY) |

BIG BANG, RAPID EXPANSION ←——— **ERA OF SLOWING** ———→ **ERA OF ACCELERATION** ————————→

Source: Dr. Robert P. Kirshner, Harvard-Smithsonian Center for Astrophysics

The New York Times

survey teams found that the universe was accelerating instead. This pointed to the existence of some kind of dark energy permeating all of space.

For the current research, astronomers observe what are called Type Ia supernovas, stellar explosions that at their peak are brighter than a billion stars like the Sun. They are thus visible across billions of light-years of space, and a close examination of their light reveals the distances, motions and other evidence of conditions. As the light travels to Earth, the wavelengths are stretched by an amount that reflects the universe's expansion when the star exploded.

Dr. Kirshner said the four extremely distant supernovas indicated that the universe seven billion years ago was "in fact winning this sort of cosmic tug-of-war," but now dark energy is more dominant.

Scientists said they assumed that with the stretching out of space the proportion of dark energy to dark matter had been reversed. In the earlier and denser universe, matter of all kinds, the invisible dark matter and the visible ordinary matter of stars and planets, predominated.

The team of Dr. Tonry and Dr. Kirshner estimates that about 60 percent of the universe is filled with dark energy and 30 percent of the mass is dark matter. The remaining 10 percent consists of ordinary matter, only 1 percent of which is visible in the galaxies. Theorists offer roughly the same estimates and surmise that the changeover from dark matter to dark energy domination probably occurred before 6.3 billion years ago.

Dr. Perlmutter said that much more research would be necessary to determine whether the changing density of the expanding universe was the only reason dark energy came to rule cosmic dynamics. Or have the physical properties of dark energy, whatever it is, changed?

Dr. Perlmutter said that in the words of Dr. Edward Witten, a theoretical astrophysicist at the Institute for Advanced Study at Princeton, the true nature of dark energy "would be No. 1 on my list of things to figure out."

The research teams are planning new observations of more distant supernovas to determine when cosmic acceleration began and to gather clues about the properties of dark energy. Some observations will be conducted with ground-based telescopes, others with the Hubble Space Telescope. Dr. Perlmutter's group has proposed putting a spacecraft in orbit with telescopes especially designed for supernova hunting and pinning down the nature of dark energy.

In "The Extravagant Universe," published last fall by Princeton University Press, Dr. Kirshner wrote: "We are not made of the type of particles that make up most of the matter in the universe, and we have no idea yet how to sense directly the dark energy that determines the fate of the universe. If Copernicus taught us the lesson that we are not at the center of things, our present picture of the universe rubs it in."

IN REVIEW

1. What evidence is being used in attempts to measure the changing rate of expansion of the universe?

2. Why was the discovery of acceleration surprising?

3. The title of the article refers to a "stop-and-go universe." What does this mean in the context of the article?

4. What evidence supports the idea of a "stop-and-go universe"?

5. What do we mean by dark energy, and what do scientists now know (or not know) about it?

Galaxies, including our Milky Way Galaxy, are thought to grow by accumulating the stars, gas, and dark matter of other galaxies. Even though the universe as a whole is expanding, galaxies are still (ever more slowly) coalescing into larger galaxies. From the vantage point of a telescope on Earth, it is easy to see other galaxies in the process of merging. To study our own Milky Way Galaxy is a little trickier, since we are inside of it and so we can't see all of it at once. To distinguish Milky Way stars from stars that were once part of another galaxy is not easy: we need to carefully measure stellar positions, motions, and other properties. This article reports on the recent discovery of a small galaxy merging with our Milky Way.

Milky Way And Neighbor Seen to Merge

By Warren E. Leary
The New York Times, January 10, 2006

WASHINGTON, Jan. 9 — A previously unrecognized galaxy appears to be merging with the Milky Way, bringing hundreds of thousands of stars into our home galaxy that no one has noticed until now, astronomers said Monday.

A survey of the northern sky has detected a huge and diffuse structure within the confines of the Milky Way that does not seem to fit in with other parts of the galaxy that contains our solar system.

Robert H. Lupton of Princeton University told a meeting of the American Astronomical Society that the large, faint collection of stars rises almost perpendicular to the flat, spiral disk of the Milky Way. The most likely interpretation of the structure, the astronomer said, is that it is a dwarf galaxy that has been merging with our galaxy.

The dwarf galaxy lies in the direction of the constellation Virgo at an estimated distance of 30,000 light years from Earth, researchers reported. While some of the stars of the companion galaxy may have been observed with telescopes for centuries, they said, no one realized they belonged to another body because they were so close and commingled with Milky Way stars.

The Milky Way is a flat, pinwheel galaxy measuring more than 100,000 light years across and containing an estimated 200 billion stars, including the sun, along its long, spiral arms and around the bulge at its center. A light year is the distance light travels in a vacuum in a year's time, about six trillion miles.

Astronomers found the merging galaxy through the Sloan Digital Sky Survey, a project that for more than five years has been mapping the distance and characteristics of millions of objects in space. The project is operated by a consortium of universities and other institutions, and uses a telescope at Apache Point Observatory in New Mexico.

The latest study, which has been submitted to The Astrophysical Journal for publication, shows that the Milky Way is still changing and evolving, said Mario Juric, a Princeton graduate student who is the principal author of the report. "It looks as though the Milky Way is still growing, by cannibalizing smaller galaxies that fall into it," Mr. Juric said in a statement.

Reporting at the same meeting, another group of researchers said they had an explanation for a mysterious warp in the disk of the Milky Way that has baffled scientists for decades.

Leo Blitz, professor of astronomy at the University of California, and his colleagues Evan Levine and Carl Heiles charted the warp and found evidence that it is a ripple or vibration set up by two small galaxies that circle the Milky Way. These satellite galaxies, called the Magellanic Clouds, cause vibrations at certain frequencies as they pass though the edges of the Milky Way, the researchers said.

It was previously believed, Dr. Blitz said, that the Magellanic Clouds, with their combined mass being only 2 percent that of the Milky Way, were too small to influence their neighboring galaxy. However, he said, when the Milky Way's dark matter is taken into account, the motion of the small galaxies can create a wake that influences the larger one.

Dark matter, invisible material that accounts for most of the universe's mass, is 20 times more massive in the Milky Way than all visible material, including stars. According to a computer model created with Martin Weinberg, an astronomy theorist at the University of Massachusetts, Amherst, dark matter spreading from the Milky Way disk with the gas layer can enhance the gravitational influence of the Magellanic Clouds as they pass through it.

IN REVIEW

1. What evidence does Dr. Lupton cite to support his conclusion that he is observing stars that are actually members of another galaxy?

2. The article states that the merging galaxy is about 30,000 light-years from Earth. How does this distance compare to the overall size of the disk of the Milky Way? Put the relative sizes into context.

3. About how many stars is the dwarf galaxy contributing to the Milky Way? What is the ratio between the number of stars contributed by the dwarf and the total number of stars in the Milky Way? Briefly discuss the implications of your answer.

4. Given what we know about the Sun itself and its orbit around the center of the Galaxy, does it seem possible that our own Sun is a recent addition to the Milky Way? Why or why not?

Supermassive black holes are thought to reside at the centers of many galaxies. What would happen if a star passed too close to one? Recent observations may be telling us the answer.

Black Holes' Vast Power Is Documented

By John Noble Wilford
The New York Times, **February 19, 2004**

New X-ray observations by orbiting satellites have given astronomers their first telling evidence that appears to confirm what had been only theory: that a star is doomed if it ventures too close to a supermassive black hole.

The National Aeronautics and Space Administration and the European Space Agency announced yesterday the detection of a brilliant flare of X-rays from the heart of a distant galaxy, followed by a fading afterglow.

After analysis, an international team of scientists concluded that the telescopes had witnessed the overpowering gravity of a black hole as it tore apart a star and gobbled up a hearty share of its gaseous mass.

It was an act of cosmic mayhem known as a stellar tidal disruption. It removed any lingering doubt, astronomers said, that the reputation of black holes as star-destroyers is fully deserved.

"Stars can survive being stretched a small amount, but this star was stretched beyond its breaking point," said Stefanie Komossa of the Max Planck Institute for Extraterrestrial Physics in Garching, Germany, who led the discovery team. "This unlucky star just wandered into the wrong neighborhood."

Astronomers said they suspected that the ill-fated star was thrown off course by a close encounter with another star. Then it fell under the gravitational influence of a black hole and its enormous tidal forces, nothing so benign as the Moon's tug on Earth's oceans. The black hole, in effect, reached out and squeezed and stretched the star until it disintegrated.

In a televised news briefing at NASA in Washington, Günther Hasinger, also an astrophysicist at the Max Planck Institute, said, "For the first time, we really are convinced that we are seeing a star being ripped apart by a black hole."

Alex Filippenko, an astronomy professor at the University of California, Berkeley, who was not involved in the research, agreed.

"This is really fantastic," Dr. Filippenko said. "This is very strong evidence that stars are being ripped apart by supermassive black holes."

The astronomers estimated that about 1 percent of the victimized star's mass was consumed by the black hole. This small amount, they said, was consistent with theoretical predictions that the momentum and energy of star-destruction process would cause most of the star's gas to be flung away from the black hole.

The initial outburst of high-energy radiation was detected in 1992 by the German Rosat X-ray spacecraft, but the observations were fuzzy. At the time, puzzled scientists could only speculate about the nature of the flare or its origin.

The more definitive observations were made three years ago by two orbiting X-ray telescopes, NASA's Chandra and the European XMM-Newton satellites. They determined that the X-rays were coming from the center of a galaxy, RX J1242-11. The most awesome black holes, with densely packed masses equivalent to millions or billions of Suns, are found at galactic cores. This one is estimated to have a mass of about 100 million Suns.

More detailed examinations of the energy spectrum of the X-rays by the European satellite, astronomers said, revealed physical conditions similar to the expected surroundings of black holes, ruling out other possible astronomical explanations. The energy liberated by the tidal disruption was reported to be equivalent to that of a supernova, an exploding star.

Dr. Filippenko said the findings should advance understanding of black holes and provide a critical framework for theoretical models of how they grow and evolve over time.

Black holes cannot be observed directly; their gravity is so strong that nothing, not even light, can escape their clutches. What is known about black holes is mostly deduced from observations of the whirlpool motions of gas and dust and stars under their gravitational influence.

The X-ray discovery may give scientists another means of identifying the presence of black holes and learning more about their behavior.

Kim Weaver, an astrophysicist at NASA's Goddard Space Flight Center in Greenbelt, Md., said the high-energy flares of a star's disintegration by black-hole gravity could serve as flashlights to illuminate the otherwise obscuring gas and dust and to provide glimpses into the inner regions of galaxies.

The astronomers gave assurances that the Sun is far enough away from a suspected black hole at the center of the Milky Way galaxy to be well out of danger.

IN REVIEW

1. What parts of the electromagnetic spectrum are used to study events taking place around black holes? Why do the observatories for these studies need to be located in space?

2. What are tidal forces? (Hint: See Chapter 5.) How would tidal forces affect a star that came close to a supermassive black hole?

3. What evidence suggests that we have observed the radiation produced by the destruction of a star near a supermassive black hole?

4. How do these observations help scientists understand how black holes evolve?

New observations of ripples moving through a galaxy cluster are shedding new light on how supermassive black holes can influence their surroundings out to very large distances. These ripples move like sound waves, and their nature provides the key clues.

Music of the Heavens Turns Out to Sound a Lot Like a B Flat

By Dennis Overbye
The New York Times, **September 16, 2003**

Astronomers say they have heard the sound of a black hole singing. And what it is singing, and perhaps has been singing for more than two billion years, they say, is B flat—a B flat 57 octaves lower than middle C.

The "notes" appear as pressure waves roiling and spreading as a result of outbursts from a supermassive black hole through a hot thin gas that fills the Perseus cluster of galaxies, 250 million light-years distant. They are 30,000 light-years across and have a period of oscillation of 10 million years. By comparison, the deepest, lowest notes that humans can hear have a period of about one-twentieth of a second.

The black hole is playing "the lowest note in the universe," said Dr. Andrew Fabian, an X-ray astronomer at the Institute for Astronomy at Cambridge University in England.

Dr. Fabian was the leader on an international team that used NASA's Chandra X-ray Observatory to detect the black hole's notes as ripples of luminosity in the X-ray glow of the cluster. The discovery, announced last week at NASA headquarters in Washington and in a paper in the journal Monthly Notices of Royal Astronomical Society, might help solve longstanding problems regarding the structure of galaxy clusters, the largest, most massive objects in the universe, and the evolution of galaxies within them, astronomers said.

Far from being "just an interesting form of black hole acoustics," as Dr. Steven Allen of the Institute of Astronomy said in a news release, the sound waves might be the key to figuring out how such clusters grow.

Black holes, as decreed by Einstein's general theory of relativity, are objects so dense that neither light nor anything else, including sound, can escape them. But long before any sort of material disappeared into a black hole, theorists have surmised, it would be accelerated to near-light speeds by the hole's gravitational field and heated to millions of degrees as it swirled in a dense doughnut around the gates of doom, sparking X-rays and shock waves and squeezing jets of energy and particles across space.

Such furiously feeding black holes are thought to be the engines responsible for the violent quasars and other phenomena in the cores of galaxies. The new work suggests that such black holes can exert influences far beyond their host galaxies.

The biggest clusters, like the one in Perseus, can contain thousands of galaxies and trillions of stars. Paradoxically, most of the ordinary matter in them resides not in stars, but in intergalactic gas that has been heated by the fall into the cluster to temperatures of 50 million degrees or so. The gas glows brightly enough in X-rays to be seen far across the universe. Cosmologists use this X-ray glow to find clusters in the deep of spacetime.

It has long been a puzzle what keeps the cluster gas hot. Without a continuing input of energy, the gas at the center would radiate its heat, lowering its pressure, and cooler gas would flow in from the outskirts, providing fresh fuel to make stars.

"We should see stars forming in central galaxies," said Dr. Kimberly Weaver, an astronomer at the Goddard Space Flight Center in Greenbelt, Md. But they do not, she said.

Astronomers, Dr. Fabian said, suspected that black holes in the central galaxies of clusters might be keeping the cluster gas hot, but the astronomers did not know how.

As the brightest X-ray cluster in the sky, radiating 1,000 times as much energy in X-rays as visible light, Perseus is a logical laboratory for investigating the problem, Dr. Fabian explained. A particularly massive black hole is believed to lurk in a galaxy known as NGC 1275, which lies at the center of the cluster.

Two jets of radio energy shooting out of the galaxy's nucleus have blown two bubbles in the gas in the center of the cluster. In an X-ray image from the Chandra satellite released three years ago, these bubbles looked like the eyeholes of a giant eerie orange skull.

Last year, however, Dr. Fabian and his colleagues obtained a new long-exposure Chandra image of the Perseus cluster, which showed waves moving outward like ripples on a pond from the central bubbles.

The waves, they realized, might be the ideal missing link between the jets and the surrounding gas. Dr. Fabian compared the process to a child's blowing bubbles in a glass of water through a straw. In this case, the jets are the straw. The bubbles pushing against the enormous pressure of the gas surrounding them create sound waves

A One-Note Maelstrom

X-RAY JET

A supermassive black hole in the Perseus galaxy cluster is emitting sound waves 30,000 light-years across — a cosmic hum in B flat.

X-ray jets from the center of the hole create bubbles of high energy that push aside gases and make the giant waves.

BLACK HOLE

BUBBLES

SOUND WAVES

X-ray view of the Perseus cluster (left), and enhanced to show sound waves.

Sources: Chandra X-Ray Center, Harvard-Smithsonian Center for Astrophysics; NASA

The New York Times; simulation image at top by Chandra X-Ray Center/A.Hobart; others by NASA

moving out through the cluster's gas, pumping energy into it and heating it.

Other astronomers called the results beautiful, but said more study was needed to confirm that the wave process could be supplying the missing energy to the cluster. "I think it might be," said Dr. Simon White of the Max Planck Institute for Astrophysics in Garching, Germany.

The energies are as prodigious as the symphony is boring. It takes the energy of 100 million supernova explosions to blow a central bubble in the cluster. If the black hole blows such bubbles continuously and it is this energy that has been keeping Perseus hot, then the black hole in Perseus must have been playing a steady B flat for a long time, said Dr. Fabian. "It's the longest-lasting symphony we know of," said Dr. Bruce Margon, an astronomer at the Space Telescope Science Institute.

IN REVIEW

1. How do the sound waves discussed in the article differ from those detectable by the human ear?

2. What is the Perseus cluster? How did the Chandra X-ray Observatory detect sound waves in the cluster?

3. How do scientists think the sound waves are linked to a supermassive black hole? Where is the black hole located?

4. How might the energy propagating through the cluster explain the X-ray glow from the cluster's gas?

5. How does understanding this cluster help scientists to understand the evolution of galaxies more generally?

As discussed in the book and in other articles, recent astronomical observations suggest that the rate of universal expansion of space is accelerating with time. The initial evidence for this acceleration came from supernova observations. The Chandra X-ray Observatory has now added another line of evidence pointing toward an accelerating universe.

By X-Raying Galaxies, Researchers Offer New Evidence of Rapidly Expanding Universe

By Dennis Overbye
The New York Times, May 19, 2004

Correction Appended

Observations of giant clouds of galaxies far out in space and time have revealed new evidence that some mysterious force began to push the cosmos apart six billion years ago, astronomers said yesterday.

The results constitute striking confirmation of one of the weirdest discoveries of modern science: that the expansion of the universe seems to be accelerating, the galaxies flying apart faster and faster with time, under the influence of some antigravitational force. The work, astronomers said, opens up a powerful new way of investigating the nature of this "dark energy" and its effect on the destiny of the cosmos.

The astronomers used an orbiting X-ray satellite called Chandra to observe hot gases in the distant galactic clusters. By analyzing the X-rays emitted by those gases, they could calculate the distance from Earth and the speed of each of the clusters and thus trace the history of the expansion of the universe over the last 10 billion years, they said.

"The universe is accelerating," said Dr. Steve Allen of Cambridge University in England, leader of the international team that did the work. "We have found strong new evidence for dark energy."

They announced their results at a news conference at NASA headquarters in Washington. A paper describing the work has been submitted to the journal Monthly Notices of the Royal Astronomical Society.

Other astronomers hailed the X-ray cluster method as a potential

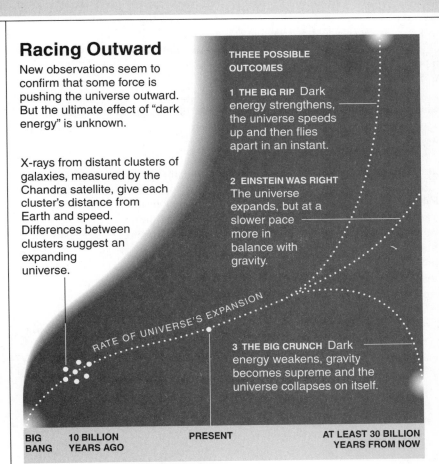

Racing Outward

New observations seem to confirm that some force is pushing the universe outward. But the ultimate effect of "dark energy" is unknown.

X-rays from distant clusters of galaxies, measured by the Chandra satellite, give each cluster's distance from Earth and speed. Differences between clusters suggest an expanding universe.

THREE POSSIBLE OUTCOMES

1 THE BIG RIP Dark energy strengthens, the universe speeds up and then flies apart in an instant.

2 EINSTEIN WAS RIGHT The universe expands, but at a slower pace more in balance with gravity.

3 THE BIG CRUNCH Dark energy weakens, gravity becomes supreme and the universe collapses on itself.

RATE OF UNIVERSE'S EXPANSION

| BIG BANG | 10 BILLION YEARS AGO | PRESENT | AT LEAST 30 BILLION YEARS FROM NOW |

Source: Harvard-Smithsonian Center for Astrophysics

The New York Times

complement to other ways of investigating dark energy but said they would withhold judgment about this particular calculation until they could study the details. Most of the previous studies, including those that led to the discovery of dark energy, used exploding stars known as Type 1a supernovas as cosmic distance markers.

Dr. Adam Riess of the Space Telescope Science Institute in Baltimore, an original discoverer of dark energy, hailed the work as another sign of the new age of "precision cosmology."

Dr. Riess said in an e-mail message: "Cosmologists are all from Missouri, the Show-Me State. It appears that X-ray clusters have been added as a new tool in our surveyor's tool kit. All tests point to a strange form of gravity we call dark energy. Some love it, some hate it; it appears we have to deal with it."

Dr. Martin Rees, a cosmologist at Cambridge who was not part of the team, called the results "neat work and a promising method," which, he noted, involved "very straightforward assumptions and simple physics."

Another cosmologist who was not part of the team, Dr. Michael Turner of the University of Chicago, said: "We can now be quite confident that the expansion of the universe is speeding up. It's not a fluke, it's not going away."

Dark energy has confounded experts and everybody else since two competing groups of astronomers discovered six years ago that the expansion of the universe was not slowing down due to cosmic gravity, as had been presumed, but was speeding up.

At the news conference, Dr. Andrew Fabian of Cambridge, a team member, compared the phenomenon to tossing an apple in the air and watching it go up faster and faster rather than falling back down. "It requires new physics beyond everyday experience, even the experience of an astronomer," Dr. Fabian said.

In recent years theorists have filled the journals with ever more fanciful explanations of what might be causing this behavior.

One possibility, first suggested and then rejected by Einstein, is that space itself has a repulsive force. But according to modern particle physics theory, this cosmological constant, as this force is known, should be about 1,060 times bigger than what astronomers have measured, causing theorists seek other explanations. Among them have

been extra unseen dimensions to space, interactions with other, parallel universes and as-yet-undiscovered particles or forces.

Or perhaps, some theorists say, Einstein's theory of gravity, general relativity, which has been the backbone of cosmology for nearly a century, needs modification.

Astronomers hope that some answers will come if they can find out whether the density of dark energy—estimated to make up 75 percent of the universe—is changing with time.

If dark energy were constant, it would mean that Einstein's cosmological constant is in effect, and that most of the galaxies would move away too fast to be seen a mere 100 billion years from now.

If dark energy is increasing, it could mean the universe could end in a "big rip," in which even atoms would be torn apart. On the other hand, the dark energy could decrease and even turn into an attractive force, drawing the universe to an end in a "big crunch."

The new results are consistent with Einstein's cosmological constant but also allow for the possibility that the dark energy could be changing, echoing recent results from the supernova surveys.

"The nice thing is that this is a completely independent method based on very simple physics," Dr. Allen said. "It's the physics of hot gas and the physics of gravity."

Clusters of galaxies are the largest objects in the universe, containing thousands of galaxies and trillions of stars. But in a big cluster, the stars themselves

are greatly outweighed by intergalactic gas, which has been condensed and heated to 100 million degrees or so by the cluster's immense gravity.

The X-rays that are spit out by this gas can be seen far across the universe. From their brightness astronomers can gauge the amount of gas in the cluster, and from the temperature of the gas, they can estimate the total mass in the galaxy cluster. Most of that mass is mysterious dark matter, which has been detected only by its gravitational effects on the luminous parts of the universe.

The astronomers made what they said was the simple assumption that clusters were a fair sample of the universe as a whole and that the cosmic ratio of dark matter to ordinary matter applied in each individual cluster. That allowed them to calculate distances to 26 clusters, from 1 billion to 10 billion light-years away, and thus measure how fast the universe was expanding when the light left those faraway galaxy clouds, confirming the cosmic acceleration.

"It's nice our results agree with previous experiments," Dr. Allen said. "It lets you feel rather more secure that everything is as it should be in those experiments."

Correction: May 21, 2004, Friday. Because of a transmission error, an article on Wednesday about a force that appears to be pushing the universe apart misstated the difference between the repulsive force as predicted by modern physics and the force actually measured. It is 1060 times as big, not 1,060 times.

IN REVIEW

1. How do observations made by the Chandra X-ray Observatory support the idea that universal expansion is accelerating?

2. Why are some astronomers reserving judgment about the observations made by Chandra?

3. Describe why dark energy is a frustrating concept for scientists. What are some of the possible explanations for dark energy?

4. How does the discussion on dark energy illustrate the hallmarks of science? In your opinion, what is the value of determining the true nature of dark energy?

How will the universe end? In simple terms, it seems that it should either continue to expand forever or someday stop expanding and begin to collapse. However, the idea of an accelerating expansion introduces new possibilities, including the possibility that the universe will eventually expand so rapidly that all matter will be torn apart in a "Big Rip."

From Space, a New View of Doomsday

By Dennis Overbye
The New York Times, **February 17, 2004**

Once upon a time, if you wanted to talk about the end of the universe you had a choice, as Robert Frost put it, between fire and ice.

Either the universe would collapse under its own weight one day, in a fiery "big crunch," or the galaxies, now flying outward from each other, would go on coasting outward forever, forever slowing, but never stopping while the cosmos grew darker and darker, colder and colder, as the stars gradually burned out like tired bulbs.

Now there is the Big Rip.

Recent astronomical measurements, scientists say, cannot rule out the possibility that in a few billion years a mysterious force permeating space-time will be strong enough to blow everything apart, shred rocks, animals, molecules and finally even atoms in a last seemingly mad instant of cosmic self-abnegation.

"In some ways it sounds more like science fiction than fact," said Dr. Robert Caldwell, a Dartmouth physicist who described this apocalyptic possibility in a paper with Dr. Marc Kamionkowski and Dr. Nevin Weinberg, from the California Institute of Technology, last year.

The Big Rip is only one of a constellation of doomsday possibilities resulting from the discovery by two teams of astronomers six years ago that a mysterious force called dark energy seems to be wrenching the universe apart.

Instead of slowing down from cosmic gravity, as cosmologists had presumed for a century, the galaxies started speeding up about five billion years ago, like a driver hitting the gas pedal after passing a tollbooth.

Dark energy sounded crazy at the time, but in the intervening years a cascade of observations have strengthened the case that something truly weird is going on in the sky. It has a name, but that belies the fact that nobody really knows what dark energy is.

In six years it has become one of the central and apparently unavoidable features of the cosmos, the surprise question mark at the top of everybody's list, undermining what physicists presumed they understood about space, time, gravity and the future of the universe.

"In five years we've gone from saying it looks like a mistake to something that everyone is claiming evidence for," said Dr. Robert Kirshner of the Harvard-Smithsonian Center for Astrophysics, who was part of the original discovery.

Dr. Saul Perlmutter, a physicist from the Lawrence Berkeley Laboratory who was a leader of one of the 1998 teams, said he thought astronomers had even gotten comfortable with the idea—"or as comfortable as you can be with something as bizarre as dark energy."

Now, armies of astronomers are fanning out into the night, enlisting telescopes, large and small, from Chile to Hawaii to Arizona to outer space, in a quest to take the measure of dark energy by tracing the history of the universe with unprecedented precision.

Some of them are following the trail blazed by the first two groups six years ago, searching out a kind of exploding star known as Type 1a the supernova. Those stars serve as markers in space, enabling scientists to plumb the size of the universe and how it grew over time. Where the first groups based their conclusions on observing a few dozen supernovas, the new efforts intend to harvest hundreds or thousands of them.

Others are seeking to gain leverage by investigating how the antigravitational force of dark energy has retarded the growth of conglomerations of matter like galaxies. In one ambitious project, a team led by Dr. John Carlstrom of the University of Chicago is building an array of radio telescopes at the South Pole to count and study clusters of galaxies deep in space-time. Others are already probing the internal dynamics of galaxies by the thousands, or building giant cameras that use the light-bending powers of gravity itself as lenses to map invisible dark matter in space and compile a growth chart of cosmic structures.

Dr. Anthony Tyson, now at Bell Laboratories, is head of a project to build a "dark matter telescope" known as the Large Synoptic Survey Telescope. "High energy physicists have been marching into our project," he said. "This is not just another telescope. It's a physics experiment, like a particle accelerator."

After all, the fate of the universe is at stake. If the dark energy is virulent enough, then that fate "is quite fantastic and completely different than the possibilities previously discussed," Dr. Caldwell and his colleagues wrote last year.

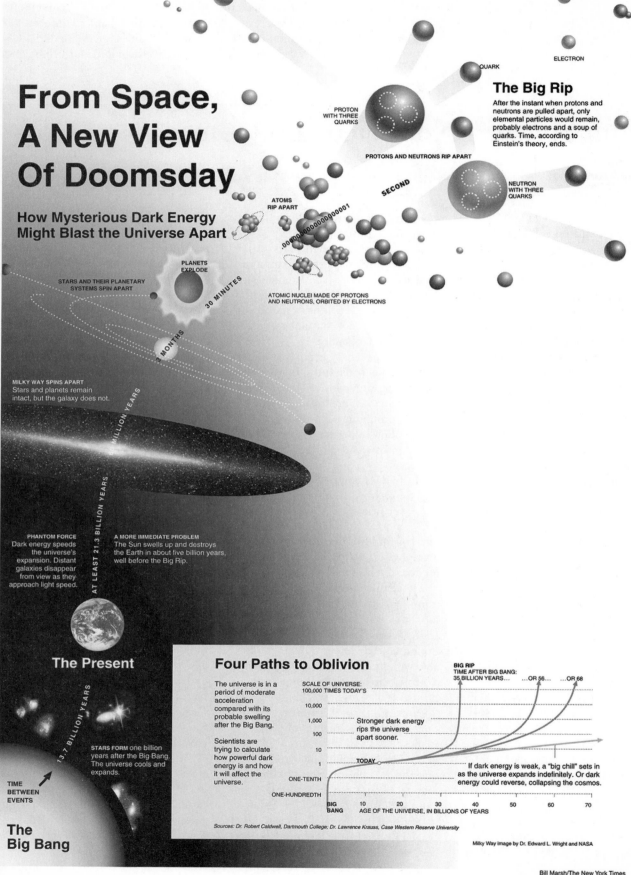

From Space, A New View Of Doomsday

How Mysterious Dark Energy Might Blast the Universe Apart

The Big Rip

After the instant when protons and neutrons are pulled apart, only elemental particles would remain, probably electrons and a soup of quarks. Time, according to Einstein's theory, ends.

ELECTRON

QUARK

PROTON WITH THREE QUARKS

PROTONS AND NEUTRONS RIP APART

SECOND

NEUTRON WITH THREE QUARKS

ATOMS RIP APART

.0000000000000000001

ATOMIC NUCLEI MADE OF PROTONS AND NEUTRONS, ORBITED BY ELECTRONS

PLANETS EXPLODE

30 MINUTES

STARS AND THEIR PLANETARY SYSTEMS SPIN APART

3 MONTHS

MILKY WAY SPINS APART
Stars and planets remain intact, but the galaxy does not.

60 MILLION YEARS

AT LEAST 21.3 BILLION YEARS

PHANTOM FORCE
Dark energy speeds the universe's expansion. Distant galaxies disappear from view as they approach light speed.

A MORE IMMEDIATE PROBLEM
The Sun swells up and destroys the Earth in about five billion years, well before the Big Rip.

The Present

13.7 BILLION YEARS

STARS FORM one billion years after the Big Bang. The universe cools and expands.

TIME BETWEEN EVENTS

The Big Bang

Four Paths to Oblivion

The universe is in a period of moderate acceleration compared with its probable swelling after the Big Bang.

Scientists are trying to calculate how powerful dark energy is and how it will affect the universe.

SCALE OF UNIVERSE: 100,000 TIMES TODAY'S

BIG RIP
TIME AFTER BIG BANG:
35 BILLION YEARS... ...OR 56... ...OR 68

Stronger dark energy rips the universe apart sooner.

10,000	
1,000	
100	
10	
1	TODAY
ONE-TENTH	
ONE-HUNDREDTH	

If dark energy is weak, a "big chill" sets in as the universe expands indefinitely. Or dark energy could reverse, collapsing the cosmos.

BIG BANG   10   20   30   40   50   60   70
AGE OF THE UNIVERSE, IN BILLIONS OF YEARS

Sources: Dr. Robert Caldwell, Dartmouth College; Dr. Lawrence Krauss, Case Western Reserve University

Milky Way image by Dr. Edward L. Wright and NASA

Bill Marsh/The New York Times

The Search for One Number

The idea of an antigravitational force pervading the cosmos does sound like science fiction, but theorists have long known that certain energy fields would exert negative pressure that would in turn, according to Einstein's equations, produce negative gravity. Indeed, some kind of brief and violent antigravitational boost, called inflation, is thought by theorists to have fueled the Big Bang.

As they try to figure out how this strange behavior could be happening to the universe today, astronomers say the ultimate prize from all the new observing projects could be as simple as a single number.

That number, known as w, is the ratio between the pressure and density of dark energy. Knowing this number and how it changes with time—if it does— might help scientists pick through different explanations of dark energy and thus the future of the universe— "whether it's gonna lead to a Big Rip, a Big Collapse or just a Big Fizzle," as Dr. Adam Riess of the Space Telescope Science Institute in Baltimore put it in an e-mail message.

One possible explanation for dark energy, perhaps the sentimental favorite among astronomers, is a force known as the cosmological constant, caused by the energy residing in empty space. It was first postulated back by Einstein in 1917. A universe under its influence would accelerate forever.

While the density of energy in space would remain the same over the eons, as the universe grows there would be more space and thus more repulsion. Within a few billion years, most galaxies would be moving away from our own faster than the speed of light and so would disappear from the sky; the edge of the observable universe would shrink around our descendants like a black hole.

But attempts to calculate the cosmological constant using the most high-powered modern theories of gravity and particle physics result in numbers 1060 times as great as the dark energy astronomers have observed— big enough, in fact, to have blown the universe apart in the first second, long before even atoms had time for form. Theorists admit they are at a loss. Perhaps, some of them now say, Einstein's theory of gravity, the general theory of relativity, needs to be modified.

Another possibility comes from string theory, the putative theory of everything, which allows that space could be laced with other energy fields, associated with particles or forces as yet undiscovered. Those fields, collectively called quintessence, could have an antigravity effect. Quintessence could change with time—for example, getting weaker and eventually disappearing as the universe expanded and diluted the field—or could even change from a repulsive force to an attractive one, which could set off a big crunch.

Recently, in a variation on the quintessence idea, Dr. Leonard Parker of the University of Wisconsin at Milwaukee and various colleagues, including Dr. Caldwell, have suggested that the field associated with some unknown very light particle could get tangled up with gravity and cause the universe to accelerate. That would alter Einstein's equations, said Dr. Caldwell. He added, "Our calculations show, however, that galaxies reside in a bubble of old-fashioned Einstein gravity, whereas gravity has changed outside and between galaxies."

A Weird Idea Gets Weirder

But the strangest notion is what Dr. Caldwell has called phantom energy, the dark energy that could lead to the Big Rip.

"It's weird negative pressure," said Dr. Lawrence M. Krauss, an astrophysicist at Case Western Reserve University in Cleveland.

While the density of the energy in Einstein's cosmological constant stays the same as the universe expands, the density of phantom energy would go up and up, eventually becoming infinite. Such would be the case if the parameter w turned out to be less than minus 1, say physicists, who admit they are stunned by the possibility and until recently simply refused to consider it.

"It crosses a boundary of good taste," Dr. Caldwell said, calling phantom energy "bad news stuff." Phantom energy violates physicists' intuitions about how the universe should behave. A chunk of it could be used to prop open wormholes in space and time—and thus create time machines, for example.

"It could lead to such bizarre effects as negative kinetic energy," Dr. Krauss said. As a result, objects like atoms would be able to lose energy by speeding up.

Nevertheless, a recent analysis by Dr. Caldwell and his Dartmouth colleague Dr. Michael Doran of the supernova measurements to date, combined with other cosmological data, suggest that w could lie anywhere from minus 0.8 to minus 1.25, leaving open the possibility of phantom energy. The cosmological constant would give a value of minus 1.0, and anything higher would be a sign of quintessence.

Dr. Kirshner said phantom energy had been dismissed as "too strange" when his group was doing calculations of dark energy back in 1998. In retrospect, he said, that was not the right thing to do.

"It sounds wacky," he said, referring to phantom energy, "but I think we're in a situation where we're going to need a really new idea. We're in trouble; the way out is going to be new imaginative things. It might be our ideas are not wild enough, they don't question fundamentals enough."

Dr. Chris Pritchet of the University of Victoria, who is part of a collaboration using the Canada-France-Hawaii telescope on Mauna Kea to search for supernovas, said, "In many ways phantom energy is unphysical, but we're not ruling it out."

Counting Down to the Big Rip

This version of doomsday would start slowly. Then, billions of years from now, as phantom energy increased its push and the cosmic expansion accelerated, more and more galaxies would start to disappear from the sky as their speeds reached the speed of light.

But things would not stop there. Some billions of years from now, depending on the exact value of w, the phantom force from the phantom energy will be enough to overcome gravity and break up clusters of galaxies. That will happen about a billion years before the Big Rip itself.

After that the apocalypse speeds up. About 900 million years later, about 60 million years before the end, our own Milky Way galaxy will be torn apart. Three months before the rip, the solar system will fly apart. The Earth will explode when there is half an hour left on the cosmic clock.

The last item on Dr. Caldwell's doomsday agenda is the dissolution of atoms, 10-19, a tenth of a billionth of a billionth of a second before the Big Rip ends everything.

"After the rip is like before the Big Bang," Dr. Caldwell said. "General relativity says: "The end. Time can't evolve."

The cosmos probably still has a lot of life in it, according to recent calculations by Dr. Krauss. Based on the current age of the universe, some 14 billion years, and other data, w cannot be less than about minus 1.2, he said, putting the Big Rip about 55 billion years in the future.

"It can't be very phantom," Dr. Krauss said.

The dark energy surveys now under way hope to be able to measure w to an accuracy of 5 percent, but even if that can be done, it may not be sufficient to eliminate the nightmare of phantom energy.

"It's hard to measure anything in astronomy to a few percent," said Dr. Sandra Faber of the University of California at Santa Cruz, who directs one of the dark energy surveys. Variations in the atmosphere and gaps in astronomers' understanding of supernova explosions add uncertainty to the dark energy measurements.

As a result some astronomers fear that the results may leave us on the razor's edge unable to decide between a cosmological constant and the other possibilities—quintessence or a Big Rip. Cosmologists could then be stuck with a "standard model" of the universe that fits all the data, but which they have no hope of understanding.

If the parameter w comes out to be something other than minus 1, Dr. Krauss said, it will at least give some direction to physicists.

One encouraging sign—"a tantalizing bit of hope," in Dr. Krauss's words— that the data will distinguish between a cosmological constant and the other possibilities came last fall when Dr. Riess, of Baltimore, reported, based on new observations of distant supernovas, that the "cosmic jerk" when dark energy took over the universe happened only five billion years ago.

In the standard cosmological constant model, said Dr. Riess, the turnaround should have come one or two billion years earlier.

Dr. Tyson was more sanguine. "Dark energy is crazy, right?" he said. "It's going to be exciting no matter what we find."

Dark Future for Dark Energy?

The work of the dark energy hunters has been complicated by the impending loss of the Hubble Space Telescope, which can see far enough out in space and time to measure how and if the dark energy parameter w is changing over the eons.

Last month, citing safety, NASA canceled all future shuttle maintenance missions to the telescope, dooming it to die in orbit, probably within three years, according to astronomers. "The Hubble shutdown will slow us all down," Dr. Perlmutter said.

At the same time, as a result of the agency's presidentially ordered shift toward the Moon and Mars, plans for a special satellite that was to have been jointly sponsored by NASA and the Department of Energy have at least temporarily disappeared from NASA's five-year budget plan.

Dr. Perlmutter, who has devoted much of his time in the last six years to the proposed satellite experiment, said he hoped that a way would be found to keep the project on track to be launched in the next decade.

"When you have the most exciting scientific problem of the day, you don't want to wait around," he said.

IN REVIEW

1. Briefly describe the possible fate of the universe that scientists are calling the Big Rip.

2. Where does the idea of the Big Rip come from? What other possible fates for the universe are consistent with current data?

3. What do scientists mean by a "cosmological constant"? How is it related to "dark energy"?

4. Why are scientists having such a difficult time explaining the observed acceleration of the cosmos?

5. In your opinion, how close are we to knowing the eventual fate of the universe? Do you think we will ever know it with confidence? Defend your opinion.

The prior article discussed the possibility that matter throughout the universe will eventually be torn apart in a "Big Rip." This article discusses additional data shedding light on possible fates of the universe.

New Data on 2 Doomsday Ideas, Big Rip vs. Big Crunch

By James Glanz
The New York Times, **February 21, 2004**

A dark unseen energy is steadily pushing the universe apart, just as Einstein predicted, suggesting the universe may have a more peaceful end than recent theories envision, according to striking new measurements of distant exploding stars by the orbiting Hubble Space Telescope.

The energy, whose source remains unknown, was named the cosmological constant by Einstein. In a prediction in 1917 that he later called "my greatest blunder," Einstein posited a kind of antigravity force that was pushing galaxies apart with a strength that did not change over billions of years of cosmic history.

Theorists seeking to explain the mysterious force have suggested that it could, in fact, become stronger or weaker over time—either finally tearing the universe apart in a violent event called "the big rip" or shutting down in the distant future—tens of billions of years from now.

If the force somehow shut down, gravity would again predominate in the cosmos, and the universe would collapse on itself. That version of oblivion is sometimes called "the big crunch."

On Friday, Dr. Adam Riess of the Space Telescope Science Institute in Baltimore presented the first broad set of observational figures that gauge the strength of the antigravity force over time. The information, he said, suggests that the cosmos will gradually expand, cool and darken, more akin to a slide into senescence rather than a violent apocalypse.

Dr. Riess and his team, which included Drs. Louis Strolger of the science institute and Alexei V. Filippenko of the

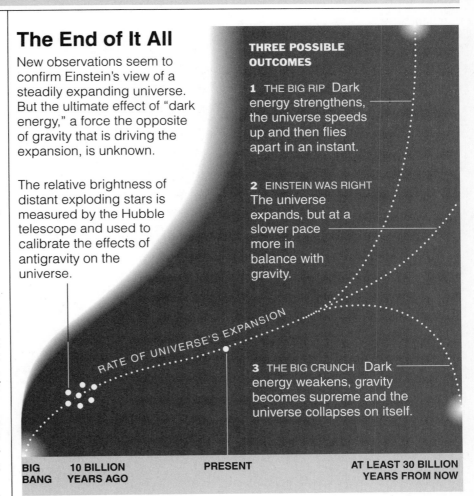

The End of It All

New observations seem to confirm Einstein's view of a steadily expanding universe. But the ultimate effect of "dark energy," a force the opposite of gravity that is driving the expansion, is unknown.

The relative brightness of distant exploding stars is measured by the Hubble telescope and used to calibrate the effects of antigravity on the universe.

THREE POSSIBLE OUTCOMES

1 THE BIG RIP Dark energy strengthens, the universe speeds up and then flies apart in an instant.

2 EINSTEIN WAS RIGHT The universe expands, but at a slower pace more in balance with gravity.

3 THE BIG CRUNCH Dark energy weakens, gravity becomes supreme and the universe collapses on itself.

RATE OF UNIVERSE'S EXPANSION

BIG BANG | **10 BILLION YEARS AGO** | **PRESENT** | **AT LEAST 30 BILLION YEARS FROM NOW**

Source: Dr. Adam Riess, Space Telescope Science Institute   The New York Times

University of California at Berkeley, used the Hubble to search for exploding stars, or supernovae, that are swept up in the cosmic push of the dark energy. They discovered 42 new supernovas in their survey area, including six of the seven most distant known.

Rather than seeing the changes in the push that many theorists had predicted, Einstein's unchanging cosmo-logical constant fits the data better than any of the alternatives.

"What we've found is that it looks like a semipermanent kind of dark energy," Dr. Riess said. "It appears like it's been with us for a long time. If it is changing, it's doing so slowly. Einstein's theory is looking a lot better than before this data."

A cosmologist not involved in the work, Dr. Michael S. Turner of the

University of Chicago, said: "This is the biggest mystery in all of science, whether or not dark energy varies with time. It's a big, big clue, and this is the first information we have."

Although the new results favor Einstein's nearly century-old prediction, Dr. Turner said, they still do not rule out some alternative theories. The information specifically leaves open the chance that the antigravity force will eventually strengthen and tear apart planets, stars and even atoms in a big rip, among other exotic possibilities.

Dr. Turner said that future measurements were quite likely to turn up smaller changes in the force over time and that those subtleties could help unravel the mystery.

The results do suggest that any ultimate cataclysm could not occur for, perhaps, 30 billion years, Dr. Riess said. Several physicists said that those estimates were highly uncertain, but that the findings could lead to a wide-ranging reassessment of models that predict strange variable energy densities in space.

"Models which predict wild dark energy densities which change a lot with time don't look so good," Dr. Yun Wang, a cosmologist at the University of Oklahoma, said. "Everybody will go back to the drawing board."

Dr. Filippenko presented some of the results at a conference here on Sources and Detection of Dark Matter and Dark Energy in the Universe.

The measurements raise new questions about the NASA decision, which is under review, to let the Hubble die a slow death in space rather than try another service mission with a space shuttle. The administrator of National Aeronautics and Space Administration, Sean O'Keefe, has said a service mission would be too risky after the disaster of the Columbia space shuttle.

Dr. Riess said he disagreed with the decision to stop the Hubble, which would halt the research for years to come. He pointed out that the remarkable clarity of the work depended on the Advanced Camera for Surveys, which spacewalking astronauts installed in Hubble two years ago.

The organizer of the conference here, Dr. David B. Cline of the University of California, Los Angeles, said that in view of the team's results, he was inclined to agree with Dr. Riess's assessment.

"They really shocked everybody by showing they could do this," Dr. Cline said. "You have to say obviously it's shame that the Hubble can't continue its life."

All the scientists agreed that because the problem had become so knotty, a complete solution might have to wait for a proposed satellite, tentatively called the Supernova/Acceleration Probe, or SNAP, which is planned to observe thousands of exploding stars from space.

"It's so tantalizing and so beautiful to see this kind of data set," said Dr. Saul Perlmutter, a physicist at Lawrence Berkeley National Laboratory who is leading the SNAP team. "This kind of result is so exciting for those of us who are eager to get to that next step."

Einstein invented the cosmological constant 1917 to explain why the universe, filled with countless stars and galaxies that are attracted to one another by gravity, did not collapse on itself. He added a constant term, called lambda, to his equations of general relativity, which describe the workings of gravity and the curvature of space on large scales. Lambda provided a repulsive force to counteract gravity.

In 1929, the astronomer Edwin Hubble, for whom the space telescope is named, discovered that the universe was expanding—from the primeval explosion called the Big Bang, scientists would learn. In that picture, gravity would slow the expansion and, possibly, reverse it at a far-off time. Einstein abandoned lambda as unnecessary and called it a blunder.

Six years ago, however, two supernova groups—one led by Dr. Perlmutter and the other by several scientists, including Drs. Filippenko and Riess—found that cosmic expansion was speeding up rather than slowing. The evidence was that distant supernovas, swept up in some sort of cosmic repulsion, were farther away than they would otherwise be and therefore appeared dimmer.

Lambda was reborn, but astronomers still had little information on whether the antigravity force was truly constant. The latest results suggest, within the accuracy of the measurements, at least, that something like Einstein's cosmological constant is indeed at work.

"His greatest blunders are our greatest ideas," Dr. Sean M. Carroll, a cosmologist at the University of Chicago, said. "It is a triumph of general relativity."

Einstein's theory says nothing about why the energy of the cosmological constant, or anything else, should be filling space in the first place, Dr. Carroll said. Theorists have turned to explanations as bizarre as parallel universes that exert an influence on our own or extra unseen dimensions in the fabric of space.

Dr. Carroll said the supernova findings might raise the stock of certain ideas that have emerged from string theory, which assumes that fundamental physics is a manifestation of minuscule strings that vibrate in an 11-dimensional space. String theorists have calculated what they call cosmic "landscapes," in which empty space contains energy densities that resemble a cosmological constant.

Because dark energy has never been directly detected and all these explanations are so bizarre, a few scientists say the ultimate explanation might be something else entirely.

Others conceded that there was a bit of a letdown that nearly a century after Einstein invented the cosmological constant nary a crack in his theory had turned up.

"This gives me a sinking feeling," Dr. Max Tegmark of the University of Pennsylvania said. "My nightmare is that we're going to be forever stuck with this puzzle."

IN REVIEW

1. What evidence supports the idea that some sort of "dark energy" is acting in the universe? What does dark energy do? How does its existence affect our ideas about the fate of the universe?

2. Why did Einstein introduce the cosmological constant, lambda, into his general theory of relativity? Why did he later call his decision to include it the biggest blunder of his career? Why might it have been?

3. Why are scientists now taking Einstein's cosmological constant seriously? How might it explain the acceleration of the cosmos?

4. The article alludes to the fact that the prior NASA administrator cancelled all future plans for servicing the Hubble Space Telescope. How would the loss of the Hubble Space Telescope affect efforts to understand the acceleration of the cosmos?

5. In mid-2005, the new NASA administrator announced that NASA is taking a new look at servicing the Hubble Space Telescope. What is the current status of plans for servicing it?

As discussed in your textbook, the most widely accepted theoretical model of the Big Bang postulates that the universe underwent a period of dramatic inflation during its first fraction of a second of existence. Inflation explains some features of the universe that are otherwise mysterious, such as why the universe is so uniform on large scales, why its geometry is so nearly flat, and what generated the original density perturbations from which galaxies later grew. Our best probe of the early universe—and our only current way to test the idea of inflation—comes from detailed study of the cosmic microwave background. This article describes the deepest analysis to date of the cosmic background radiation, which has been mapped by the Wilkinson Microwave Anisotropy Probe (WMAP).

Astronomers Find the Earliest Signs Yet of a Violent Baby Universe

By Dennis Overbye
The New York Times, March 17, 2006

Using data from a new map of the baby universe, astronomers said yesterday that they had seen deep into the Big Bang, and had gotten their first detailed hint of what was going on less than a trillionth of a second after time began.

The results, they said, validated a key prediction of the speculative but popular cosmic theory known as inflation about the distribution of matter and energy in the Big Bang. The theory holds that during its first moments, the universe, fueled by an antigravitational field, underwent a violent growth spurt, ballooning from submicroscopic to astronomical size in the blink of an eye.

"It amazes me that we can say anything about the universe in the first trillionth of a second," said Charles L. Bennett, a professor at the Johns Hopkins University and the leader of the group that reported the results yesterday. "It appears that the infant universe had the kind of growth spurt that would alarm any mom or dad." The map was produced by a NASA satellite known as the Wilkinson Microwave Anisotropy Probe that has been circling the Earth at a point on the other side of the Moon since 2001, recording faint emanations of microwaves thought to be the remnants of the Big Bang.

The microwaves paint a portrait of the 13.7-billion-year-old universe when

New Evidence

By measuring patterns in the polarization of the radiation from the Big Bang (map at left), astronomers were able to refine their measurements of the infant universe. They found strong evidence that the universe had undergone a violent growth spurt in the first trillionth of a second of time.

PATTERNS OF POLARIZATION
Direction of light vibration.

A Big Bang Timeline

Astronomers are now able to more accurately calculate a timeline for major events in the formation of the early universe.

BIG BANG RADIATION

DEVELOPMENT OF GALAXIES, PLANETS, ETC.

DARK ENERGY ACCELERATED EXPANSION

INFLATION THEORY
Rapid expansion, during which small fluctuations form the seeds of galaxies.

FIRST STARS
About 400 million years.

WILKINSON SATELLITE

13.7 BILLION YEARS

Sources: NASA; Johns Hopkins University   The New York Times

it was only 380,000 years old, astronomers say. But in the details of that portrait are clues to processes that occurred when it was much younger.

Using the map, the Wilkinson team has been able to revise an earlier estimate of the time at which the first stars began to form and shine through the primordial murk that followed the cooling of the Big Bang. Those stars appeared when the universe was about 400 million years old, they said yesterday.

The previous estimate of 200 million years, based on earlier Wilkinson data, had been seen as surprisingly early by many cosmologists, and the new date is comfortably in line with mainstream theories.

Inflation theory, which was invented by Alan H. Guth of the Massachusetts Institute of Technology, has been the workhorse of Big Bang cosmology for the last 25 years. But astronomers and physicists admit that they still have no idea what caused inflation. As a result, there are a welter of models describing how it might have worked.

Although inflation is not yet conclusively confirmed, it is now in better shape than ever, many astronomers said, and many models can be eliminated.

"We've crossed a threshold," said David N. Spergel of Princeton University, a member of the research team. "We can now start to say something interesting about the physics of inflation."

Others not involved in the project tended to agree.

"If this holds up to the test of time, it's a real landmark," said Max Tegmark, a cosmologist at M.I.T.

Dr. Guth, who is at a conference in the Caribbean, was said to be walking around with a big smile.

The new map has been eagerly awaited by astronomers, who last heard from the Wilkinson group in 2003 when it released its first map. That map showed the cosmos speckled with faint hot and cool spots, the seeds of structures like galaxies.

Three years is a long time to go between baby pictures.

Dr. Bennett and his colleagues have spent the time taking a much more dif-ficult measurement, in effect using spacecraft antennas to measure the polarization of the Big Bang microwaves. To make these measurements, which required 100 times the sensitivity of the previous heat measurements, the astronomers essentially had to recalibrate their entire spacecraft and the way they looked at the data.

"We had to rewrite the whole software pipeline — twice," Dr. Spergel said. The light waves from the Big Bang, they found, do not vibrate randomly in different directions as they travel from the distant past to us. Rather, they have a slight preference to line up in one plane.

Lyman Page, a Princeton physicist, compared it to the glare of sunlight bouncing off the hood of a car. The reflection causes the light waves to oscillate in a horizontal direction. In the case of the car, a person would wear sunglasses to eliminate the glare. In the case of the Big Bang microwaves, he said, "We measure the glare."

What plays the role of the hood of the car in this story, Dr. Page said, is a fog of electrons floating in space between the Earth and the Big Bang. This fog was produced, so the story goes, by radiation from the first stars ripping apart atoms in space and liberating their electrons.

Measuring the polarization more accurately allowed the Wilkinson team to refine its previous estimate of when the stars first turned on.

In turn, by correcting for the effects of this electron fog, the astronomers were able to measure fluctuations in the microwaves more accurately than they had before.

This allowed them to confront for the first time an important prediction of inflation theory. For 30 years, cosmologists had presumed that the waves and ripples in the early universe followed a simple pattern, namely that their brightness was independent of their size. But according to inflation theory, the brightness of the bumps should be slightly dependent on their apparent sizes in the sky. Smaller bumps should be slightly dimmer than big ones.

The reason, explained Dr. Spergel, is that the force driving inflation is falling as it proceeds. The smaller bumps would be produced later and so a little less forcefully than the bigger ones.

That, in fact, is what the Wilkinson probe has measured. Dr. Spergel said, "It's very consistent with the simplest inflation models, just what inflation models say we should see."

Michael Turner, a cosmologist at the University of Chicago, called the results, "the first smoking gun evidence for inflation."

But Paul Steinhardt, a physics professor at Princeton who has lately championed an alternative to inflation in which the universe begins and ends cyclically in a collision between two island universes, pointed out that the new data are also compatible with his theory.

Andrei Linde, a Stanford University physicist, noted that his favorite model was still in the running and exclaimed in an e-mail message, "Great day for cosmology!"

The stage is set for a race to achieve the next milestone, proof of inflation theory. If it is correct, there should be a separate, even fainter pattern of polarization, because of gravity waves, the roiling of space-time by the violent wrench of inflation.

Detecting those waves would confirm inflation. Dr. Spergel said that if those waves were there the Wilkinson probe might eventually be able to see them. Next year, the European Space Agency is scheduled to launch its Planck satellite, which will also search for gravitational waves. In the meantime, balloon- and ground-based telescopes will also take aim at the cosmic microwaves.

Dr. Steinhardt said, "If you want to know where you came from, and where you're going, that's the issue at stake."

IN REVIEW

1. When we map the cosmic background radiation, what are we actually mapping? Why can such maps reveal characteristics of the universe at very young ages—even back to a fraction of a second after the Big Bang—even though the radiation itself was released much later?

2. What is polarization? Give an example of polarized light from everyday experience.

3. Why has it taken longer to analyze the WMAP data for polarization, and how does this new analysis teach us more about the early universe?

4. What do the new data tell us about when the first stars formed? Why were astronomers somewhat relieved by this new time estimate?

5. Do the new data prove that inflation really occurred? Defend your opinion.

Physicists have long thought that nature should be inherently simple. The current picture, in which four forces govern interactions among many types of particles, seems too complex. About 30 years ago, scientists successfully showed that two of the four forces—the electromagnetic and weak forces—merge together into a single electroweak force at high energies. But the "holy grail" of physics is finding a single theory that explains all four forces and the diversity of particles. String theory attempts to do this, and progress in string theory is summarized in this article.

String Theory, at 20, Explains It All (or Not)

By Dennis Overbye
The New York Times, December 7, 2004

They all laughed 20 years ago. It was then that a physicist named John Schwarz jumped up on the stage during a cabaret at the physics center here and began babbling about having discovered a theory that could explain everything. By prearrangement men in white suits swooped in and carried away Dr. Schwarz, then a little-known researcher at the California Institute of Technology.

Only a few of the laughing audience members knew that Dr. Schwarz was not entirely joking. He and his collaborator, Dr. Michael Green, now at Cambridge University, had just finished a calculation that would change the way physics was done. They had shown that it was possible for the first time to write down a single equation that could explain all the laws of physics, all the forces of nature—the proverbial "theory of everything" that could be written on a T-shirt.

And so emerged into the limelight a strange new concept of nature, called string theory, so named because it depicts the basic constituents of the universe as tiny wriggling strings, not point particles.

"That was our first public announcement," Dr. Schwarz said recently.

By uniting all the forces, string theory had the potential of achieving the goal that Einstein sought without success for half his life and that has embodied the dreams of every physicist since then. If true, it could be used like a searchlight to illuminate some of the deepest mysteries physicists can imagine, like the origin of space and time in the Big Bang and the putative death of space and time at the infinitely dense centers of black holes.

In the last 20 years, string theory has become a major branch of physics. Physicists and mathematicians conversant in strings are courted and recruited like star quarterbacks by universities eager to establish their research credentials. String theory has been celebrated and explained in bestselling books like "The Elegant Universe," by Dr. Brian Greene, a physicist at Columbia University, and even on popular television shows.

Last summer in Aspen, Dr. Schwarz and Dr. Green (of Cambridge) cut a cake decorated with "20th Anniversary of the First Revolution Started in Aspen," as they and other theorists celebrated the anniversary of their big breakthrough. But even as they ate cake and drank wine, the string theorists admitted that after 20 years, they still did not know how to test string theory, or even what it meant.

As a result, the goal of explaining all the features of the modern world is as far away as ever, they say. And some physicists outside the string theory camp are growing restive. At another meeting, at the Aspen Institute for Humanities, only a few days before the string commemoration, Dr. Lawrence Krauss, a cosmologist at Case Western Reserve University in Cleveland, called string theory "a colossal failure."

String theorists agree that it has been a long, strange trip, but they still have faith that they will complete the journey.

"Twenty years ago no one would have correctly predicted how string theory has since developed," said Dr. Andrew Strominger of Harvard. "There is disappointment that despite all our efforts, experimental verification or disproof still seems far away. On the other hand, the depth and beauty of the subject, and the way it has reached out, influenced and connected other areas of physics and mathematics, is beyond the wildest imaginations of 20 years ago."

In a way, the story of string theory and of the physicists who have followed its siren song for two decades is like a novel that begins with the classic "what if?"

What if the basic constituents of nature and matter were not little points, as had been presumed since the time of the Greeks? What if the seeds of reality were rather teeny tiny wiggly little bits of string? And what appear to be different particles like electrons and quarks merely correspond to different ways for the strings to vibrate, different notes on God's guitar?

It sounds simple, but that small change led physicists into a mathematical labyrinth, in which they describe themselves as wandering, "exploring almost like experimentalists," in the words of Dr. David Gross of the Kavli Institute for Theoretical Physics in Santa Barbara, Calif.

String theory, the Italian physicist Dr. Daniele Amati once said, was a piece of 21st-century physics that had fallen by accident into the 20th century.

And, so the joke went, would require 22nd-century mathematics to solve.

Dr. Edward Witten of the Institute for Advanced Study in Princeton, N.J., described it this way: "String theory is not like anything else ever discovered. It is an incredible panoply of ideas about math and physics, so vast, so rich you could say almost anything about it."

The string revolution had its roots in a quixotic effort in the 1970's to understand the so-called "strong" force that binds quarks into particles like protons and neutrons. Why were individual quarks never seen in nature? Perhaps because they were on the ends of strings, said physicists, following up on work by Dr. Gabriele Veneziano of CERN, the European research consortium.

That would explain why you cannot have a single quark—you cannot have a string with only one end. Strings seduced many physicists with their mathematical elegance, but they had some problems, like requiring 26 dimensions and a plethora of mysterious particles that did not seem to have anything to do with quarks or the strong force.

When accelerator experiments supported an alternative theory of quark behavior known as quantum chromodynamics, most physicists consigned strings to the dustbin of history.

But some theorists thought the mathematics of strings was too beautiful to die.

In 1974 Dr. Schwarz and Dr. Joel Scherk from the École Normale Supérieure in France noticed that one of the mysterious particles predicted by string theory had the properties predicted for the graviton, the particle that would be responsible for transmitting gravity in a quantum theory of gravity, if such a theory existed.

Without even trying, they realized, string theory had crossed the biggest gulf in physics. Physicists had been stuck for decades trying to reconcile the quirky rules known as quantum mechanics, which govern atomic behavior, with Einstein's general theory of relativity, which describes how gravity shapes the cosmos.

That meant that if string theory was right, it was not just a theory of the strong force; it was a theory of all forces.

"I was immediately convinced this was worth devoting my life to," Dr. Schwarz recalled "It's been my life work ever since."

It was another 10 years before Dr. Schwarz and Dr. Green (Dr. Scherk died in 1980) finally hit pay dirt. They showed that it was possible to write down a string theory of everything that was not only mathematically consistent but also free of certain absurdities, like the violation of cause and effect, that had plagued earlier quantum gravity calculations.

In the summer and fall of 1984, as word of the achievement spread, physicists around the world left what they were doing and stormed their blackboards, visions of the Einsteinian grail of a unified theory dancing in their heads.

"Although much work remains to be done there seem to be no insuperable obstacles to deriving all of known physics," one set of physicists, known as the Princeton string quartet, wrote about a particularly promising model known as heterotic strings. (The quartet consisted of Dr. Gross; Dr. Jeffrey Harvey and Dr. Emil Martinec, both at the University of Chicago; and Dr. Ryan M. Rohm, now at the University of North Carolina.)

The Music of Strings

String theory is certainly one of the most musical explanations ever offered for nature, but it is not for the untrained ear. For one thing, the modern version of the theory decreed that there are 10 dimensions of space and time.

To explain to ordinary mortals why the world appears to have only four dimensions—one of time and three of space—string theorists adopted a notion first bruited by the German mathematicians Theodor Kaluza and Oskar Klein in 1926. The extra six dimensions, they said, go around in sub-submicroscopic loops, so tiny that people cannot see them or store old National Geographics in them.

A simple example, the story goes, is a garden hose. Seen from afar, it is a simple line across the grass, but up close it has a circular cross section. An ant on the hose can go around it as well as travel along its length. To envision

the world as seen by string theory, one only has to imagine a tiny, tiny six-dimensional ball at every point in space-time.

But that was only the beginning. In 1995, Dr. Witten showed that what had been five different versions of string theory seemed to be related. He argued that they were all different manifestations of a shadowy, as-yet-undefined entity he called "M theory," with "M" standing for mother, matrix, magic, mystery, membrane or even murky.

In M-theory, the universe has 11 dimensions—10 of space and one of time, and it consists not just of strings but also of more extended membranes of various dimension, known generically as "branes."

This new theory has liberated the imaginations of cosmologists. Our own universe, some theorists suggest, may be a four-dimensional brane floating in some higher-dimensional space, like a bubble in a fish tank, perhaps with other branes—parallel universes—nearby. Collisions or other interactions between the branes might have touched off the Big Bang that started our own cosmic clock ticking or could produce the dark energy that now seems to be accelerating the expansion of the universe, they say.

Toting Up the Scorecard

One of string theory's biggest triumphs has come in the study of black holes. In Einstein's general relativity, these objects are bottomless pits in space-time, voraciously swallowing everything, even light, that gets too close, but in string theory they are a dense tangle of strings and membranes.

In a prodigious calculation in 1995, Dr. Strominger and Dr. Cumrun Vafa, both of Harvard, were able to calculate the information content of a black hole, matching a famous result obtained by Dr. Stephen Hawking of Cambridge University using more indirect means in 1973. Their calculation is viewed by many people as the most important result yet in string theory, Dr. Greene said.

Another success, Dr. Greene and others said, was the discovery that the shape, or topology, of space, is not fixed but can change, according to string theory. Space can even rip and tear.

But the scorecard is mixed when it comes to other areas of physics. So far, for example, string theory has had little to say about what might have happened at the instant of the Big Bang.

Moreover, the theory seems to have too many solutions. One of the biggest dreams that physicists had for the so-called theory of everything was that it would specify a unique prescription of nature, one in which God had no choice, as Einstein once put it, about details like the number of dimensions or the relative masses of elementary particles.

But recently theorists have estimated that there could be at least 10100 different solutions to the string equations, corresponding to different ways of folding up the extra dimensions and filling them with fields—gazillions of different possible universes.

Some theorists, including Dr. Witten, hold fast to the Einsteinian dream, hoping that a unique answer to the string equations will emerge when they finally figure out what all this 21st-century physics is trying to tell them about the world.

But that day is still far away.

"We don't know what the deep principle in string theory is," Dr. Witten said.

For most of the 20th century, progress in particle physics was driven by the search for symmetries—patterns or relationships that remain the same when we swap left for right, travel across the galaxy or imagine running time in reverse.

For years physicists have looked for the origins of string theory in some sort of deep and esoteric symmetry, but string theory has turned out to be weirder than that.

Recently it has painted a picture of nature as a kind of hologram. In the holographic images often seen on bank cards, the illusion of three dimensions is created on a two-dimensional surface. Likewise string theory suggests that in nature all the information about what is happening inside some volume of space is somehow encoded on its outer boundary, according to work by several theorists, including Dr. Juan Maldacena of the Institute for Advanced Study and Dr. Raphael Bousso of the University of California, Berkeley.

Just how and why a three-dimensional reality can spring from just two dimensions, or four dimensions can unfold from three, is as baffling to people like Dr. Witten as it probably is to someone reading about it in a newspaper.

In effect, as Dr. Witten put it, an extra dimension of space can mysteriously appear out of "nothing."

The lesson, he said, may be that time and space are only illusions or approximations, emerging somehow from something more primitive and fundamental about nature, the way protons and neutrons are built of quarks.

The real secret of string theory, he said, will probably not be new symmetries, but rather a novel prescription for constructing space-time.

"It's a new aspect of the theory," Dr. Witten said. "Whether we are getting closer to the deep principle, I don't know."

As he put it in a talk in October, "It's plausible that we will someday understand string theory."

Tangled in Strings

Critics of string theory, meanwhile, have been keeping their own scorecard. The most glaring omission is the lack of any experimental evidence for strings or even a single experimental prediction that could prove string theory wrong—the acid test of the scientific process.

Strings are generally presumed to be so small that "stringy" effects should show up only when particles are smashed together at prohibitive energies, roughly 1019 billion electron volts. That is orders of magnitude beyond the capability of any particle accelerator that will ever be built on earth. Dr. Harvey of Chicago said he sometimes woke up thinking, What am I doing spending my whole career on something that can't be tested experimentally?

This disparity between theoretical speculation and testable reality has led some critics to suggest that string theory is as much philosophy as science, and that it has diverted the attention and energy of a generation of physicists from other perhaps more worthy pursuits. Others say the theory itself is still too vague and that some promising ideas have not been proved rigorously enough yet.

Dr. Krauss said, "We bemoan the fact that Einstein spent the last 30 years of his life on a fruitless quest, but we think it's fine if a thousand theorists spend 30 years of their prime on the same quest."

The Other Quantum Gravity

String theory's biggest triumph is still its first one, unifying Einstein's lordly gravity that curves the cosmos and the quantum pinball game of chance that lives inside it.

"Whatever else it is or is not," Dr. Harvey said in Aspen, "string theory is a theory of quantum gravity that gives sensible answers."

That is no small success, but it may not be unique.

String theory has a host of lesser known rivals for the mantle of quantum gravity, in particular a concept called, loop quantum gravity, which arose from work by Dr. Abhay Ashtekar of Penn State and has been carried forward by Dr. Carlo Rovelli of the University of Marseille and Dr. Lee Smolin of the Perimeter Institute for Theoretical Physics in Waterloo, Ontario, among others.

Unlike string theory, loop gravity makes no pretensions toward being a theory of everything. It is only a theory of gravity, space and time, arising from the applications of quantum principles to the equations of Einstein's general relativity. The adherents of string theory and of loop gravity have a kind of Microsoft-Apple kind of rivalry, with the former garnering a vast majority of university jobs and publicity.

Dr. Witten said that string theory had a tendency to absorb the ideas of its critics and rivals. This could happen with loop gravity. Dr. Vafa; his Harvard colleagues, Dr. Sergei Gukov and Dr. Andrew Neitzke; and Dr. Robbert Dijkgraaf of the University of

Amsterdam report in a recent paper that they have found a connection between simplified versions of string and loop gravity.

"If it exists," Dr. Vafa said of loop gravity, "it should be part of string theory."

Looking for a Cosmic Connection

Some theorists have bent their energies recently toward investigating models in which strings could make an observable mark on the sky or in experiments in particle accelerators.

"They all require us to be lucky," said Dr. Joe Polchinski of the Kavli Institute.

For example the thrashing about of strings in the early moments of time could leave fine lumps in a haze of radio waves filling the sky and thought to be the remains of the Big Bang. These might be detectable by the Planck satellite being built by the European Space Agency for a 2007 launching date, said Dr. Greene.

According to some models, Dr. Polchinski has suggested, some strings could be stretched from their normal submicroscopic lengths to become as big as galaxies or more during a brief cosmic spurt known as inflation, thought to have happened a fraction of a second after the universe was born.

If everything works out, he said, there will be loops of string in the sky as big as galaxies. Other strings could stretch all the way across the observable universe. The strings, under enormous tension and moving near the speed of light, would wiggle and snap, rippling space-time like a tablecloth with gravitational waves.

"It would be like a whip hundreds of light-years long," Dr. Polchinski said.

The signal from these snapping strings, if they exist, should be detectable by the Laser Interferometer Gravitational Wave Observatory, which began science observations two years ago, operated by a multinational collaboration led by Caltech and the Massachusetts Institute of Technology.

Another chance for a clue will come in 2007 when the Large Hadron Collider is turned on at CERN in Geneva and starts colliding protons with seven trillion volts of energy apiece. In one version of the theory—admittedly a long shot—such collisions could create black holes or particles disappearing into the hidden dimensions.

Everybody's favorite candidate for what the collider will find is a phenomenon called supersymmetry, which is crucial to string theory. It posits the existence of a whole set of ghostlike elementary particles yet to be discovered. Theorists say they have reason to believe that the lightest of these particles, which have fanciful names like photinos, squarks and selectrons, should have a mass-energy within the range of the collider.

String theory naturally incorporates supersymmetry, but so do many other theories. Its discovery would not clinch the case for strings, but even Dr. Krauss of Case Western admits that the existence of supersymmetry would be a boon for string theory.

And what if supersymmetric particles are not discovered at the new collider? Their absence would strain the faith, a bit, but few theorists say they would give up.

"It would certainly be a big blow to our chances of understanding string theory in the near future," Dr. Witten said.

Beginnings and Endings

At the end of the Aspen celebration talk turned to the prospect of verification of string theory. Summing up the long march toward acceptance of the theory, Dr. Stephen Shenker, a pioneer string theorist at Stanford, quoted Winston Churchill:

"This is not the end, not even the beginning of the end, but perhaps it is the end of the beginning."

Dr. Shenker said it would be great to find out that string theory was right.

From the audience Dr. Greene piped up, "Wouldn't it be great either way?"

"Are you kidding me, Brian?" Dr. Shenker responded. "How many years have you sweated on this?"

But if string theory is wrong, Dr. Greene argued, wouldn't it be good to know so physics could move on? "Don't you want to know?" he asked.

Dr. Shenker amended his remarks. "It would be great to have an answer," he said, adding, "It would be even better if it's the right one."

IN REVIEW

1. What is the goal of string theory? What is the basic idea behind it?

2. Describe a few of the successes of string theory over the past 20 years.

3. Why is it so difficult to put string theory to the test? Given that difficulty, is string theory really a "theory" by our usual scientific definition? Explain.

4. Why are some scientists skeptical of whether string theory is even on the right track, let alone of its being the right theory to explain the nature of the universe?

5. Based on what you've read, how much effort do you think scientists should continue to make in pursuing string theory? Defend your opinion.

The movie *Contact* opens with a scene in which the character played by Jodie Foster is shown patiently listening through headphones amidst giant radio telescopes. The character is not listening to music, but to static that may contain some faint signal from intelligent life beyond Earth. Although the movie is science fiction, it is based on actual work being done by the SETI Institute, a private, nonprofit organization whose mission is to promote deeper understanding of the potential for life beyond Earth. The SETI Institute is not focused entirely on searching for extraterrestrial intelligent life; institute scientists also study the origin of life on Earth and the potential for life of any kind on other worlds.

Search for Life Out There Gains Respect, Bit by Bit

By Dennis Overbye
The New York Times, July 8, 2003

Years after Congress ordered NASA to pull the plug on a survey looking for alien radio signals from the stars, the Search for Extraterrestrial Intelligence, or SETI, as it is known to aficionados, seems to have gradually achieved a modicum of respect in the halls of Washington.

The most recent indication appeared at the end of last month, when NASA named 12 groups that had won five-year grants to participate as "lead teams" in its Astrobiology Institute, which investigates the origin and future of life in the universe.

On the list was the SETI Institute, an organization in Mountain View, Calif., that has carried on the abandoned survey.

The group proposed a variety of basic research on the way planetary environments affect life or are affected by it. One project is aimed at determining whether certain kinds of stars are promising abodes for life and thus good targets for a planned expansion of the institute's search for intelligent radio signals.

That would make the grant the first money in a decade that NASA has allocated for work related to radio searches, the astronomers at the institute said.

For most of the last decade, "SETI was a four-letter word in NASA," said Dr. Frank Drake, a radio astronomer and former chairman of the SETI Institute. "It was not uttered in speeches, or in documents."

SETI Institute

A radio telescope in Green Bank, W. Va., is part of SETI's network.

NASA said nothing had changed. The agency does not as a rule finance ground-based astronomy and, thus, has no SETI (pronounced SEH-tee) program. But SETI research can be supported as long as it meets the strictures of good science and emerges from a competitive peer-reviewed process, explained Dr. Edward J. Weiler, associate administrator of the NASA Office of Space Science.

Dr. Michael Meyer, a biologist who heads the Astrobiology Institute, described the proposed study as "pure astronomy," aimed at looking for potential habitable planets, research that fits with the institute's mission. NASA, he added, is eager to use the results to find targets for its planned Terrestrial Planet Finder satellite.

Astronomers around SETI and elsewhere said NASA and Congress had recently shown warming attitudes toward the politically embattled subject of intelligent life Out There. Dr. Martin Rees, a cosmologist at the University of Cambridge in England, used an e-mail message to attribute the change partly to growing scientific interest in extraterrestrial biology and the origins of life, as well as, perhaps, "the growing visibility and manifest professionalism of the SETI Institute."

The issue is so important, Dr. Rees said, that "even though the chances of success are exceedingly low, it's worth a moderate effort."

Dr. Michael Turner, an astrophysicist at the University of Chicago, used a football analogy to describe the odds, saying, "SETI is definitely throwing deep, and oh what a touchdown it would be!"

Astronomers who testified two years ago to a House subcommittee on space and aeronautics reported that members seemed to support SETI. According to the journal Nature, Representative Lamar Smith, the Texas Republican who called the hearing, said the discovery of life elsewhere in the universe would be "one of the most astounding discoveries in human history."

"Funding should match public interest," Mr. Smith said, "and I don't believe it does."

In recent years, the reports from the National Academy of Sciences have endorsed the idea of SETI, and the institute itself, once advised to change its name, has become a respected "brand" in astrobiology, said Dr. Drake, as evidenced by the recent announcement. "All of this is indeed a major sea change," he said.

It was Dr. Drake who in 1960 first pointed a radio telescope at two stars, hoping to hear the cosmic equivalent of "hi there." He did not hear anything, but he earned a curious sort of scientific immortality for his frustration.

No amount of cosmic silence has been able to discourage astronomers who theorize that radio signals can bridge the unbridgeable gulfs between stars much more cheaply than spacecraft, allowing distant species to communicate by a sort of cosmic ham radio. The Milky Way has 200 billion stars, the astronomers point out, and billions of frequencies available for signaling, if "they" exist.

SETI had been in eclipse at NASA since September 1993, when Congress, fearing a backlash if it spent tax dollars on "little green men," amended the NASA appropriation to kill a 10 year search program that had begun a year before.

Part of the project, a survey of 1,000 nearby Sun-like stars, was to have been managed by NASA's Ames Research Center in Moffett Field, Calif., and carried out by the SETI Institute, formed in 1984 as a conduit for scientists to obtain grants and conduct astrobiology research. When NASA pulled out, the astronomers at the SETI Institute decided to take the search private.

In 1995, with grants from Silicon Valley titans like David Packard, William Hewlett and Dr. Barney Oliver of Hewlett-Packard; Gordon Moore, co-founder of Intel; and Paul G. Allen, co-founder of Microsoft; as well as Arthur C. Clarke, the science fiction author and inventor of the communications satellite, the search was reborn as Project Phoenix.

The astronomers expect to finish surveying the original list in the fall, but they are already laying plans for an expanded survey of up to a million stars.

The survey will be performed by a dedicated array of 350 small radio telescopes that will be built in conjunction with the University of California at Berkeley, at its Hat Creek Observatory near Mount Lassen in Northern California. The telescope, which can be used for SETI and regular radio astronomy, will be known as the Allen Telescope Array, after Mr. Allen, who invested $11.5 million for developing the telescopes.

Meanwhile, the SETI Institute, with about 120 employees and an annual budget of $10 million, not counting the cost of the new telescope array, has grown into a powerhouse of astrobiological research.

Its scientists say they have never had any trouble obtaining support from NASA and the National Science Foundation for this branch of their activities, which are as diverse as studying the chemistry of interstellar clouds and ways to handle Martian soil samples.

In all, the institute has more than 36 individual grants and cooperative agreements with NASA, said Thomas Pierson, chairman of the institute.

"Only our SETI work is (has been, actually) without federal support for the past 10 years," Mr. Pierson wrote in an e-mail message.

That exclusion has particularly stung, as NASA has embarked on highly publicized programs to search for cosmic origins—of life, matter and everything else—and begun planning for the Terrestrial Planet Finder, which will search for Earth-like planets.

The search for intelligent life in the cosmos was a logical part of those endeavors, the staff argued, and should be eligible for federal money.

"The questions SETI asks are a natural component of the questions that get asked in astrobiology," said Dr. Christopher F. Chyba, a Stanford professor who is head of astrobiology research at the SETI Institute.

As an indication of changing fortunes, astronomers point to remarks that Dr. Weiler made at the House subcommittee hearing on July 12, 2001. Dr. Weiler made clear in the session that the Congressional ban on SETI applied just to 1994.

"NASA is no longer prohibited by any congressional language from considering funding SETI research," he said, "so SETI is currently eligible and considered fairly under peer review for NASA opportunities."

The next year, a National Academy of Sciences report, "Life in the Universe, an Assessment of U.S. and International Programs in Astrobiology," called the SETI Institute a "unique endeavor" and an "important national resource in astrobiology."

The search for intelligent life, it said, is "the most romantic and publicly accessible aspect of the search for life, yet is perhaps the most problematic."

"It would be dissembling," the report added, "to say the least, to discourage such a search (especially one enabled by private funding) at the same time that astrobiology as a whole taps into the same emotions and aspirations to excite the public about the general search for life's origins, evolution and cosmic ubiquity."

Last year, the idea of searching for intelligent signals was explicitly endorsed in the newest version of "The NASA Astrobiology Roadmap," an outline of questions and goals assembled by 200 astrobiologists inside and outside NASA as a guide to research.

Although technology may be rare in the universe, its effects may nevertheless be easier to detect from a distance than biological ones.

"Accordingly," the roadmap said, "current methods should be further developed and novel methods should be identified for detecting electromagnetic radiation or other diagnostic artifacts that indicate remote technological civilizations."

The selection of the SETI Institute as part of the Astrobiology Institute is more evidence, Dr. Chyba said, that "there has been a sea change in attitudes toward SETI, as evidenced by the 'Astrobiology Roadmap' and, explicitly, by NASA."

The higher end of evolution, intelligence, has been missing from astrobiology, he said, adding, "We're bringing it to the institute now for the first time."

In a statement from the institute, Dr. Jill Tarter, a radio astronomer who directs the search program, called the selection "the stamp of approval that what we started so long ago is a really good idea."

"The cross-fertilization of all these disciplines," Dr. Tarter said, "pays big dividends."

IN REVIEW

1. Briefly describe the funding history of the SETI Institute.

2. Why has SETI research been controversial? How have attitudes toward this research changed in recent years?

3. What is the Allen Telescope Array, and what will it be used for?

4. A Congressman is quoted in the article as saying, "Funding should match public interest." Explain what he means in the context of SETI research. Do you think his statement offers good guidance for scientific funding decisions in general? Why or why not?

5. In your opinion, should the government fund SETI research? Would your opinion change if we found microbial life on Mars? Explain.